IMAGES

UNITED STATES NAVAL AVIATION 1911-2014

RARE PHOTOGRAPHS FROM WARTIME ARCHIVES

Michael Green

Pen & Sword
AVIATION

First published in Great Britain in 2015 by
PEN & SWORD AVIATION
An imprint of
Pen & Sword Books Ltd
47 Church Street
Barnsley
South Yorkshire
S70 2AS

ISBN 978-1-47382-225-2

Typeset by Concept, Huddersfield, West Yorkshire HD4 5JL.
Printed and bound in China by Imago

Pen & Sword Books Ltd incorporates the imprints of Pen & Sword Archaeology, Atlas, Aviation, Battleground, Discovery, Family History, History, Maritime, Military, Naval, Politics, Railways, Select, Social History, Transport, True Crime, and Claymore Press, Frontline Books, Leo Cooper, Praetorian Press, Remember When, Seaforth Publishing and Wharncliffe.

For a complete list of Pen & Sword titles please contact
PEN & SWORD BOOKS LIMITED
47 Church Street, Barnsley, South Yorkshire, S70 2AS, England
E-mail: enquiries@pen-and-sword.co.uk
Website: www.pen-and-sword.co.uk

Contents

Dedication

I would like to dedicate this book to US Navy fighter pilot John (Ted) Crosby, who, flying his Grumman F6F Hellcat fighter shot down five Japanese aircraft on 16 April 1945, making him the very rare ace-in-a-day pilot. Crosby survived the war and retired as a Commander in July 1969.

Acknowledgments

As with any published work authors must often depend on their friends for assistance. These include Michael Panchyshyn and fellow author Mark Stille who took the time to review the text. The bulk of the historical photographs in this work were acquired from the National Naval Aviation Museum, and the National Archives, with additional historical photos coming from Real War Photos (operated by Jo Ellen & George D. Chizmar).

For brevity's sake the picture credits for the National Naval Aviation Museum are shortened to 'NNAM' and those from Real War Photos will be shortened to 'RWP'.

Modern post-war images that came from the US Navy News Service and Defense Imagery, which is operated by the United States Department of Defense, are shortened to 'DOD'. Other contemporary pictures were supplied by my friends Christopher Vallier and the husband and wife team of Paul and Loren Hannah.

Foreword

Few things at sea are more impressive than watching a current US Navy aircraft carrier conduct flight operations. It is a carefully choreographed dance of men, women, machines and technology. It has been honed over time to routinely include difficult evolutions such as simultaneous launch and recovery operations, and operations in all types of weather and at night. Few things in aviation are more difficult and it is daily testimony to the capabilities of the aviators, their aircraft, and the carrier crew.

Though this is the state of US naval aviation today, it came from uncertain beginnings. The initial operations of aircraft from ships were little more than publicity stunts with no practical operational applicability. Nevertheless, a dedicated group of air power advocates kept moving aviation forward in the US Navy. By continually pressing the bounds of technology, they made aviation relevant and useful in a navy which was wedded to the notion that the Big Gun was the final arbiter of naval power. Only some thirty years after the first flight from an American ship, naval aviation became dominant. What started as experimentation led to a reshaping of warfare at sea.

Initially, naval aviation was limited to land-based aircraft and a small number of floatplanes based on ships. This all began to change with the introduction of USS *Langley* in 1922. From her deck, and from the decks of the first fleet carriers, USS *Lexington* and USS *Saratoga*, appeared an array of fighters, dive-bombers and torpedo aircraft which laid the basis for a naval revolution. These primitive aircraft did not immediately threaten the battleship, but by the start of the Pacific War, they possessed the technology and tactics to change the face of naval warfare.

By the end of the Second World War, the US Navy possessed over 40,000 aircraft. In spite of the unquestioned wartime success of naval aviation, it was faced with a struggle for its survival as the Cold War looked to be a war fought between nuclear powers in which conventional combat would count for little. Thankfully, this notion was quickly shown to be false, and naval aviation proved itself in a number of conventional conflicts.

The unquestioned success of naval aviation in Korea led to the authorization of the first of a series of 'super carriers'. These would become Cold War icons, moving around the world as instruments of US national policy and protecting American interests. On a routine basis, they were used to actually project power ashore in conflicts large and small. The size and capabilities of these ships allowed them to carry a stunning array of aircraft capable of any mission and equal to the aircraft of any other air force in the world.

None of this was possible without the aircraft to make US Navy aircraft carriers into the war-winning platforms that they proved to be. This book traces the US Navy's aircraft from humble beginnings into the future when stealth and unmanned aircraft will form the basis for air power at sea. It is important to keep in mind that naval aviation is not just the aircraft and helicopters on carriers, but also land-based aircraft which play a critical role in supporting fleet operations.

Most readers will be familiar with the standard US Navy aircraft which fought and won the war in the Pacific, and probably with the best-known Cold War aircraft. But there is much more to the story of naval aviation, and this single volume covers almost all American naval land-based and float aircraft, as well as the pre-Second World War developmental aircraft which flew from carriers which are often overlooked. Most impressive is the coverage given to post-Second World War aircraft which were introduced in bewildering types and often left service quickly as technology advanced with little notice or recognition.

With his usual incisive text and supporting array of well-chosen photographs, author Michael Green has provided a useful account of the development and history of US Navy Aviation. It is a story that deserves to be told and told well which this book accomplishes.

Mark E. Stille
Commander, United States Navy (retired)

Notes to the Reader

1. This book is a broad overview of the vast subject of aviation in the US Navy, from its earliest, humble beginnings, to the trappings of a modern-day superpower. It is not intended to be a comprehensive history.

2. Many of the aircraft mentioned in this book came or come in a wide variety of models, sometimes configured for many different roles. Due to the breadth of this book's scope, not every aircraft, or every aircraft model, is shown or described.

3. The chapter breakdown is somewhat arbitrary, with many of the aircraft over-lapping multiple time periods. As a general rule, the aircraft are divided among the chapters not by their first flight, but by their introduction into service, which can be months or years later. However, there are exceptions to this, in order to maintain a sense of continuity in the chapter and caption text, for aircraft that transcend the chapter headings.

4. US Navy aircraft initially had designations associated with the company who made them. On 27 March 1914 the original company designations of naval aircraft were changed to a two-letter prefix code and a number. The first letter denoted class, the second letter the type of aircraft within a class. The number that followed the two-letter prefix code stood for the order in which the aircraft was acquired. Four classes were set up: the letter 'A' stood for heavier-than-air, 'D' stood for airships, 'B' for balloons, and 'K' for kites. Within the heavier-than-air 'A' class, the letters L, H, B, X, and C stood respectively for landplanes, floatplanes, seaplanes, amphibians, and convertibles.

5. US Navy aircraft prior to 1922 sometimes retained their builder's names and designation systems. In 1922, the US Navy began a not always consistent program of classifying aircraft with its own unique designation system, which lasted until 1962. As the US Navy has always been responsible for the procurement of aircraft for the Marine Corps, all of that service's aircraft conformed to the US Navy's 1922–1962 designation system. Many of the aircraft described in this book also saw service with the US Marine Corps.

6. In the 1922–1962 designation system, the first letter/letters stood for the type of airplane. Early examples from the 1920s and 1930s included; 'F' for fighter, 'P' for patrol, 'S' for scout, 'SC' for scout-bomber, 'O' for observation, and 'OS'

for observation-scout; although sometimes the first prefix code letter might be assigned to stand for a prototype, or experimental aircraft. A new basic mission code prefix letter designation appeared in 1946, which was 'A' for an attack plane. At the same time, the designations 'TB' for torpedo-bomber and 'SB' for scout-bomber (dive bomber) eventually vanished from use, as did the aircraft they represented before and during the Second World War.

The number/numbers following the first letter/letters of an aircraft's designation represented the purchase-order of the plane from that builder. The letter following the number stood for the company's name.

Some company code letters can be readily identifiable as the first letter in a firm's name, while others do not have that connection. An example would be the Consolidated Aircraft Corporation, which was assigned the letter 'Y'. Company code letters and the firms they represented could also change or be reassigned over the years as aircraft designers and manufacturers went out of business, were sold, or merged with other companies.

The final number in an aircraft's designation was the sequence number of that model of plane built. As an example, in 1931, the US Navy took into service the F4B-2, which translates into fighter plane/the fourth aircraft type acquired from that builder/Boeing as the builder/the second model or version of that aircraft series constructed.

During the period between the First and Second World Wars, some manufacturers' names for their products were used to identify certain aircraft, or series of aircraft. British official nicknames for American-supplied planes were adopted on occasion. These names were not officially recognized by the US Navy at the time, but are used in this work.

In late 1941, the US Government adopted the British wartime policy of assigning a general name for all models of a certain type of airplane, with an example being the 'Hellcat' fighter which came in a number of versions. This policy was continued postwar by the American military, examples being the 'Skyhawk' and 'Phantom'. The British themselves had originally assigned their own names to American supplied aircraft, however they later decided to keep the original American names.

7. In 1962, the US Navy had to change its aircraft designation system to conform to a new 'Tri-Service Aircraft Designation System' mandate. In effect, the US Navy was forced to adopt the existing US Air Force aircraft designation system.

With the new 1962 aircraft designation system, the first prefix letter was sometimes its mission modification code letter, which identified any special features of the aircraft. An example would be the letter 'E' for special electronic installation, 'R' for photo-reconnaissance, 'S' for anti-submarine, and 'M' for missile-carrying aircraft. However things do change, and 'M' now stands for multi-purpose aircraft, or helicopter.

If there was no need for a mission modification code letter, the first letter in an aircraft's designation was the basic mission code. The number/numbers following the basic mission code letter were assigned by the services based on the plane acquisition sequence since 1962. A single letter followed the number(s) and identified an aircraft's variant in alphabetical order, beginning with the letter 'A'.

As an example, in December 1970, Grumman Aerospace Corporation first flew an aircraft labelled the 'YF-14A'. The letter 'Y' meant it was a prototype. If it was the letter 'X' it would mean the aircraft was experimental. The letter 'F' stood for the aircraft type, in this case a fighter. The number following the second letter meant it was the fourteenth fighter acquired by the US Navy/US Air Force since the introduction of the Tri-Service Aircraft Designation System mandate in 1962. The last letter, following the number, meant it was the first model of the original aircraft design.

Despite the 1962 Tri-Service Aircraft Designation System mandate, to have all the services conform to the same aircraft identification system it has been some-times inconsistently applied by both the US Navy and US Air Force, with numbers skipped for different reasons, such as a cancellation of a proposed plane before it reached production, or to hide a secret (black) program.

8. The number of units built of some aircraft types can vary depending on the reference sources quoted.

Chapter One

Pre-Second World War Aircraft

The early years of US naval aviation were marked by rapid changes in technology and extremely limited budgets in the post-First World War era. That meant the US Navy bought small lots of aircraft from numerous manufacturers, always seeking the best in performance while it transitioned from biplane to monoplane, and tested tactics and doctrine for its new aircraft and carriers. Aircraft did not last long in frontline service as they rapidly were superseded by newer, more advanced designs.

The Wright brothers, Orville and Wilbur, made their first powered flight on 17 December 1903, at Kitty Hawk, North Carolina. They did not stand on their laurels and continued to refine the design of their aircraft to improve its performance. The US Navy was not that interested, but did have the foresight to send military observers to aerial demonstrations within the United States and overseas to monitor the progress of aviation technology.

The continuing advancement in aircraft designs following the Wright Brothers' demonstration in 1903 was becoming harder for the US Navy's senior leadership to ignore as the years went on. In response, the Secretary of the US Navy (a civilian-appointed position) placed Captain Washington I. Chambers in charge of all aviation matters.

To show off the potential offered by aircraft, Captain Chambers had a short temporary wooden platform built on the bow of the US Navy cruiser USS *Birmingham* (CL-2). On 14 November 1910, civilian pilot Eugene B. Ely successfully flew off the ship, in a civilian wheeled airplane built by the Curtiss Aeroplane Company, founded by Glen H. Curtiss. Approximately two weeks later, Curtiss offered to train a single US Navy officer how to fly, and was taken up on his offer.

In the meantime, Captain Chambers arranged to have a longer temporary wooden platform constructed on the stern of the US Navy cruiser USS *Pennsylvania* (ACR-4). It was intended to provide the room needed for Eugene B. Ely to land and then take off from the ship (then moored in San Francisco Bay) a feat that he accomplished on 18 January 1911. It was also the first use of arresting gear, to slow down and stop the aircraft after landing.

Glen H. Curtiss made the first successful floatplane flight, taking off from San Diego Bay on 26 January 1911, in an aircraft he designed and his firm built. The following month, Captain Chambers arranged to have Curtiss taxi his floatplane out to the USS *Pennsylvania*, now moored in San Diego Bay. Once adjacent to the ship it was hoisted aboard by a crane and then lowered back into the water. This test was conceived by Chambers and Curtiss to prove the ability of floatplanes to operate from US Navy ships.

Development Continues

The publicity tests put together by Chambers and Curtiss had the desired effect on the US Navy's senior leadership, and on 4 March 1911 the first funds for naval aviation were appropriated. It was at this point in time the Wright Brothers offered to train a single US Navy officer how to fly, if, in exchange, the service would purchase one of their aircraft for the sum of $5,000.

On 8 May 1911, Chambers prepared the necessary paperwork for the US Navy to acquire three aircraft. Of the three planes, the first two to enter service were Curtiss-designed and built aircraft. The first of these two Curtiss airplanes to arrive was nicknamed the A-1. The initial flight of the A-1, in its wheeled configuration, took place on 1 July 1911. The ability of the aircraft to land and take-off from the water as a floatplane was demonstrated ten days later. A few days later the second Curtiss aircraft ordered was delivered to the US Navy and was designated the A-2.

The US Navy also took into service the first of a small number of Wright Brothers-designed and built aircraft in July 1911. The first plane to arrive was designated the B-1. It, and the other airplanes acquired from the Wright Company (formed in 1909 by the Wright Brothers) were eventually configured as training floatplanes.

Pre-First World War Floatplanes and Flying Boats

The US Navy took into service the first of five Curtiss flying boats in 1912, which the company labelled the Model F. They were initially numbered in sequence from C-1 through C-5 by the US Navy. They received a two-letter prefix code of 'AB' in March 1914. The 'A' stood for heavier-than-air-craft and the 'B' for flying boat. AB-2 was the first US Navy aircraft launched by catapult from a ship while underway, an event that took place on 5 November 1915.

In January 1916, the Curtiss Aeroplane Company became the Curtiss Aeroplane and Motor Company. By 1917, Curtiss Model F flying boats had become the US Navy's standard training aircraft, with 150 built between 1916 and 1917.

Unlike floatplanes that depend on under-fuselage pontoons for buoyancy on the water, flying boats depend on their boat-shaped fuselages for buoyancy. Flying boats often employed outrigger pontoons on the end of their wings for added stability in the water. At the time, flying boats were known as hydro-aeroplanes. Floatplanes and flying boats both fall under the umbrella term of seaplanes.

In October 1913, the US Navy's second Curtiss-built aircraft, designated the A-2, had its original float pontoon replaced by a flying boat fuselage containing a three-wheeled landing gear. This provided the aircraft the ability to land and takeoff from airfields, or the water. With this added feature, the airplane became an amphibian flying boat. Floatplanes can also be fitted with wheels to become amphibians.

Naval Aviation in Action

The largest contribution made by naval aviation during the First World War proved to be the establishment of a number of shore bases along the French, Irish, and Italian coasts. From these bases, US Navy pilots and air crews primarily flew anti-submarine patrols in flying boats, such as the Curtiss-designed twin engine H-16. The H-16 was eventually superseded in production by the F-5L, a US-modified version of a British-designed flying boat.

When the First World War came to an end on 11 November 1918, naval aviation had 2,107 aircraft, with many of them being floatplanes of foreign design. By 1919, the bulk of the naval aviation assets acquired during the First World War were gone; scrapped or placed into storage. The thousands of men who had manned and serviced these craft were demobilized. As the First World War was also known as 'The War to End All Wars', few politicians or their civilian constituents foresaw any need for retaining the men and their equipment.

A Vision is formulated

Looking forward it was clear to the senior leadership of the US Navy that the future of naval aviation lay in having a suitable array of aircraft to perform a variety of tasks. As a result, on 13 March 1919, the Chief of Naval Operations issued a preliminary program for post-First World War development. That program called for a number of specialized aircraft for service with the fleet on its battleships and cruisers, as well as the newly-envisioned inventory of aircraft carriers, typically shortened to just carrier.

In addition to those aircraft planned for use with the fleet when at sea, the US Navy also saw a requirement for land-based patrol planes that could protect convoys from enemy submarine attacks. A hitch in that requirement occurred in 1931 when the US Navy and US Army Air Corps agreed that only the latter could operate shore-based multi-engine aircraft. This restricted the US Navy to employing shore-based seaplanes and ship-based aircraft.

Catapult-Launched Aircraft

Beginning in 1926, catapult-launched single-engine floatplanes, officially classified as scouting and observation planes, began serving on US Navy battleships and cruisers. Their main job eventually became spotting for naval gunfire, the scouting role being taken over by wheeled carrier aircraft. They also had a backup role as utility aircraft and occasionally performed in the search and rescue role. All of the catapult-launched single-engine floatplanes could be fitted with wheels in place of their large under-fuselage float if required.

The first generation of scouting and observation floatplanes launched by catapult from US Navy battleships and cruisers all came from Chance Vought, with the initial production model designated the O2U-1 Corsair. Progressively improved models were labelled the O2U-2 through O2U-4 Corsair. By 1930, another variant was referred to as the O3U Corsair, evolving from the O3U-1 through O3U-4 Corsair.

When the O3U Corsair series was assigned to carriers in the wheeled observation-only configuration, they were relabelled SU-1 through SU-4. The later production models of the planes lasted in service up to early 1942. Rather than US Navy squadrons, the Corsair observation series planes assigned to carriers where flown by Marine pilots, the first to fly from US Navy carriers. In total, 580 units of the Corsair series were built by Vought.

The letter 'U' in the designation code for the various versions of the Corsair observation planes stood for Chance-Vought. Over the decades to follow, the firm would pass through numerous iterations of corporate ownership, as did many aircraft builders. To simplify, the name Vought will be retained hereafter to identify all aircraft from the various corporate entities that owned the firm.

The eventual replacement for the early generation Vought catapult-launched single-engine floatplanes was the Curtiss SOC Seagull, of which 323 were acquired in various models. Like the floatplanes that came before, it was a biplane. It entered into US Navy service in 1935 and was eventually superseded in service by the Vought OS2U Kingfisher, beginning in August 1940. Unlike all the catapult-launched single-engine floatplanes that came before it, the Kingfisher was a monoplane. The US Navy took in 1,159 units of the Kingfisher series.

Patrol Flying Boats

To replace its aging inventory of First World War-era patrol flying boats, the Naval Aircraft Factory (NAF) decided to upgrade the design of the F5L patrol flying boat. The resulting aircraft appeared in US Navy service in 1924 as the PN-7. It was followed into service by small numbers of successively improved versions, designated the PN-8 through PN-12, with the latter being the definitive design. The letter 'P' stood for patrol and the letter 'N' for the first letter in the builder's name, in this case the government-owned NAF.

The US Navy eventually decided they wanted an improved version of the PN-12. As the NAF could not build as many improved units of the PN-12 as the US Navy desired, it contracted with various civilian firms in 1929 to build an upgraded version. The first to enter service was the Douglas PD-1. It was later joined in service by the

Martin PM-1, the Keystone PK-1, and the Hall Aluminum Company's PH-1. All were basically the same aircraft with minor differences between the various companies' products.

In the late 1920s, the US Navy was looking for the next generation of flying boats. First acquired was the twin-engine Martin P3M, which was a Consolidated Aircraft-designed aircraft for which Martin had won the production contract. However, only nine units of the Martin P3M series were built. The Martin P3M was quickly superseded by the US Navy's adoption of the superior Consolidated P2Y flying boat ordered in 1931. There were three models of the P2Y. Both the Martin P3M and the Consolidated P2Y were pulled from frontline US Navy service in late 1941.

The frontline replacement for the Consolidated P2Y and the Martin P3M was the twin-engine Consolidated PBY Catalina, which entered into US Navy service in 1936. The letter 'B' in the designation prefix code stood for 'bomber' and the letter 'Y' for Consolidated. The number of Catalina series aircraft completed by 1945 ranges from a low of 2,300 to a high of 3,100 units, depending on the reference sources quoted. Many of the Catalina series aircraft would serve into the early postwar years.

Martin delivered to the US Navy a twin-engine amphibian seaplane beginning in 1940, designated the PBM-1 Mariner. It was built in numerous models, up through the PBM-5, which served during the Second World War up through the early postwar years, including the Korean War. There were at least 1,000 units of the Mariner constructed by 1945.

1920 Fighters

During the First World War, the US Navy had borrowed some foreign-designed and built fighters from the US Army, but never employed them in combat. Immediately after the conflict, the US Navy continued to acquire US Army aircraft, one of these being the Vought VE-7 two-seat trainer. The US Navy thought so highly of the aircraft that they brought a single seat variant into service in 1920 as a fighter, designated the VE-7S.

The VE-7S was the first aircraft launched from the flight deck of the then still experimental USS Langley (CV-1) on 17 October 1922. In 1925, the first US Navy squadron assigned to the USS Langley (now in fleet service) was equipped with eighteen of the VE-7S. The fighter remained in US Navy use until 1927. For a single year in 1923, the US Navy referred to its fighters as pursuit planes, which is how the US Army Air Corps identified their fighter planes.

Carrier-Capable Fighters

The first purpose-built fighter actually ordered by the US Navy and intended for carrier use was the TS-1 that appeared in service in 1922. It was designed by a civilian working for the Bureau of Aeronautics, which had been established the year before.

The new bureau had brought almost all the once divergent responsibilities for US Navy aircraft under one roof. Curtiss won the contract to build thirty-four units of the TS-1 fighter, with the government-owned NAF assigned to build five of them. The aircraft lasted in US Navy service until 1930.

Boeing soon jumped into the competition for supplying the US Navy with the fighter planes it needed. In 1925, the US Navy took into service ten of the land-based Boeing FB-1 fighters. It was followed into service by approximately thirty FB-2 through FB-5 fighters, the latter version entering service in 1927. The Boeing FB-2, FB-3, and FB-5 had been designed for carrier use, while the FB-4 was a seaplane fighter, only one being built. The letter 'F' in the designation codes stood for fighter and the letter 'B' for Boeing.

Boeing also began building for the US Navy another series of fighters, with a different type of engine. These aircraft were all intended for carrier use and labelled F4B-1 through F4B-4, with the first delivery of the F4B-1 in 1929 and the last, the F4B-4 in 1932. The latter was the last Boeing fighter built for the US Navy before the Second World War. In total, 186 units of the Boeing F4B-1 through F4B-4 were built, and they would continue to fly with the US Navy until 1939.

Curtiss also wanted in on the US Navy contracts for fighters and achieved success in 1925 when he made his first delivery of an aircraft, designated the F6C-1 Hawk, of which the US Navy ordered nine units for carrier use. Of the nine F6C-1 Hawks ordered, four were eventually completed as the F6C-2 Hawk. Two years later he began delivery of thirty-five units of a carrier fighter referred to as the F6C-3. The aircraft could be configured as a seaplane if the need arose by the replacement of its wheels with floats. Following the F6C-3 into service were thirty-one units designated the F6C-4. The letter 'C' in the aircraft designation stood for Curtiss.

1930 Fighters

In 1929, the Curtiss Aeroplane and Motor Company merged with the Wright Aeronautical Corporation to become the Curtiss-Wright Corporation, hereafter referred to as Curtiss-Wright for the sake of brevity. The following year, Curtiss-Wright was awarded a contract by the US Navy for five examples of a multi-role carrier-based twin-seat fighter-bomber/observation plane designated the F8C-1 Falcon to be employed by the Marine Corps.

Service use quickly demonstrated the Falcon was much slower than existing single-seat fighters and it was pulled from frontline US Navy carrier use in 1931. The US Marine Corps continued to employ the aircraft as a land-based observation plane. Reflecting its new job, the F8C-1 was re-designated the OC-1, with the letter 'O' standing for observation plane. Twenty-one units of an improved model built by Curtiss-Wright were labelled the F8C-3 Falcon and re-designated the OC-2 Falcon.

A follow-on version, the F8C-4, was nicknamed the Helldiver instead of the Falcon, and twenty-five were delivered to the US Navy beginning in 1930. They were quickly turned over to naval reserve units and the US Marine Corps the following year. It was then designated as the O2C Helldiver. A follow-on F8C-5 variant, primarily employed by the US Marine Corps, was labelled as the O2C-1 Helldiver, and remained in service until 1936. The US Marine Corps saw the O2C and O2C-1 as having a secondary role as dive-bombers.

In 1932, Curtiss-Wright began delivery of twenty-eight units of another carrier fighter-bomber, designated the F11C-2 Goshawk, which was re-designated as the BFC-2 Goshawk in early 1934. The re-designation better reflected the aircraft's dual-purpose role as both a fighter and as a dive bomber. The letter 'B' in the new designation stood for bomber and the 'F' for fighter. The original Goshawk was followed into service in 1934 by another twenty-seven units of an improved version, designated BF2C-1 Goshawk. The two-letter prefix code 'BF' was a short-lived US Navy designation.

The Grumman Aircraft Engineering Corporation entered the contest for supplying fighters for the US Navy at the request of the Bureau of Aeronautics in 1931. The US Navy liked what they saw and took into service the first of twenty-seven units of a two-man aircraft in 1933, labelled the FF-1. The second 'F' in the designation code stood for Grumman. The FF-1 lasted in US Navy frontline service until approximately 1935. In a secondary role as a trainer the aircraft survived in use until 1942. In its two-seat scout configuration it was designated SF-1 and thirty-three were purchased by the US Navy.

Unlike the fighter designs of its competitors, the FF-1 had an enclosed cockpit and retractable undercarriage, which led to an increase in maximum speed as the airplane's fuselage was more streamlined. These design features led to the Grumman aircraft out-performing the fighter designs from both Curtiss and Boeing.

The FF-1 from Grumman was superseded in US Navy service in 1935 by the delivery of fifty-four units of an improved one-man version of the aircraft, designated the F2F-1. It remained in use on carriers until 1940. The F2F was then overtaken on the production line by the F3F-1 in 1936, and then through the F3F-3. The latter entered service in 1938, and remained in the inventory until 1940. A total of 162 units of the F3F series were built.

On the Eve of War Fighters

In late 1935, the US Navy opened a competition for the next generation of carrier fighters. Three companies vied for the contract. In the end, a fighter, designated the F2A-1, was chosen. The letter 'A' in the designation code stood for the Brewster Aeronautical Corporation.

The F2A-1 was the first monoplane fighter in US Navy service; all those that had come before were biplanes. The US Navy ordered fifty-four of the aircraft in 1938. However when completed, forty-three went to the Finnish Government and only eleven were taken into US Navy service at the end of 1939.

At the same time as the US Navy ordered the F2A-1 into production, it asked Grumman, one of the competing firms for the contract, to keep working on improving their own aircraft design, which was a biplane and greatly impressed the US Navy. This was done as a backup plan in case the F2A-1 did not live up to the US Navy's expectations.

The delivery of forty-two units of an improved version of the F2A-1 to the US Navy, referred to as the F2A-2, occurred in late 1940. Unfortunately, combat reports from the Royal Air Force (RAF) indicated that the F2A-2 was badly outclassed by German front-line fighters. It was the RAF that officially named the F2A-2 the 'Buffalo', a name then adopted by the US Navy for the entire series. A small number of Buffalos would go on to serve with the Royal Navy (RN).

A request for more armor protection on the F2A-2 by the US Navy and foreign users of the plane resulted in the production of the next version, designated the F2A-3. It was delivered to the US Navy a few months before the Japanese attack at Pearl Harbor. However, the plane's added weight made it difficult to fly. Service use of the heavier F2A-3 demonstrated that its landing gear system was not tough enough to withstand the repeated shock of landings upon carrier flight decks and it was quickly pulled from frontline service by the US Navy, although it saw service with the US Marine Corps during the Battle of Midway

The Brewster Buffalo Replacement

The various design issues with the Buffalo prompted the US Navy to go back to Grumman. By this time the firm had come up with a suitable monoplane fighter design, the first prototype of which had flown in September 1937 and easily outperformed the Brewster monoplane fighter. The US Navy wasted no time and ordered seventy-three units of the new Grumman fighter design in August 1939, for testing, and designated it the F4F-3.

Delivery of the F4F-3 to the US Navy began in February 1940. It had a supercharged 1,200 hp engine that gave the single-seat fighter a maximum speed of 331 mph. In March 1941, the first of ninety-five units of an improved model, labelled the F4F-3A, was delivered to the US Navy. It featured a slightly different 1,200 hp engine than the first model of the aircraft. It was also fitted with a simpler single-stage supercharger because there was a shortage of the two-stage supercharger found in the initial version of the aircraft.

Based on early British combat experience with the F4F-3, a new up-gunned and armored version was placed into production by Grumman in early 1942. The

US Navy designated it the F4F-4 and named it the Wildcat. A total of 1,169 units were built. By the middle of 1942, it had replaced the majority of the earlier variants of the plane on US Navy carriers.

To allow Grumman to concentrate on building the next-generation fighter to replace the Wildcat, the US Navy assigned production of F4F aircraft to the Eastern Aircraft Division of General Motors, hereafter referred to as General Motors for the sake of brevity.

The General Motors near-copy of the Grumman F4F-4 Wildcat was labelled the FM-1. The General Motors FM-2 model of the Wildcat was based on two Grumman prototypes labelled the XF4F-8, powered by 1,350 hp engines. According to one reputable source, General Motors built 1,600 units of the FM-1 and 4,777 units of the FM-2, by 1945. Other sources quote different numbers of the FM-1 and FM-2 built.

Biplane Dive Bombers

Dive bombing, an RAF invention from the First World War, had been taken up by the US Marine Corps following the conflict and adopted by the US Navy in 1928. Dive bombing involved aiming an aircraft at an enemy ship while in a steep dive and releasing the bomb at a relatively low altitude for maximum accuracy. The steep angle of attack made it very hard for enemy ship-board anti-aircraft guns to engage attacking dive bombers.

The first dedicated biplane dive bomber for the US Navy was a Martin-designed single-engine biplane, referred to as the BM, with twenty-eight being ordered in 1931. It came in two models, the BM-1 and the BM-2. The letter 'B' in the aircraft prefix code stood for bomber and the 'M' for Martin. There was a swing-out ordnance cradle underneath the plane's fuselages for a 1,000 lb bomb. Both versions of the BM dive bomber were pulled from US Navy frontline service in 1937.

The follow-on to the Martin BM-1 and BM-2 dive-bombers was the Great Lakes BG-1 dive bomber. The letter 'B' in the aircraft prefix code stood for bomber and the 'G' for Great Lakes Aircraft Company. Like the Martin product, the Great Lakes dive bomber was a single-engine biplane with a crew of two. The US Navy ordered the first batch in 1933 and the last in 1935. The order encompassed sixty units of the aircraft. It served in frontline US Navy carrier service from 1934 to 1938. The US Marine Corps employed the aircraft as a dive bomber until 1940.

Next in line were the Curtiss-Wright F11C-2 Goshawk and the F11C-3 Goshawk, in 1934 (already described in the text) that were considered both fighters and dive bombers. Employing fighters in a secondary role as dive bombers was eventually seen as comprising their primary job, so it was decided in 1934 to go back to employing dedicated dive bombers once again.

When the US Navy made the decision to field dedicated dive bombers in 1934, it also decided to assign these new specialized dive bombers a secondary role as scout planes. These two combined roles were now defined by assigning the letter prefix code 'SB' for scout-bomber. Hereafter in the text the term dive-bombers will be referred to as scout-bombers.

Biplane Scout-Bombers

A dedicated scout-bomber taken into service pre-war by the US Navy was the Vought SBU-1, which entered service in late 1935, with eighty-four units ordered. It had originally been designed for the US Navy as a two seat fighter, but as the service then wanted only single-seat fighters it was rejected. Forty units of an improved dedicated scout-bomber model designated the SBU-2 were delivered by Vought to the US Navy in 1937. Both aircraft were also referred to by the company as the Corsair, a name they would also use for many other aircraft.

The US Navy also sought out other firms to build dedicated scout-bombers. This included Curtiss-Wright who delivered in 1937 eighty-three units of an aircraft designated SBC-3. The last pre-war Curtiss-Wright dedicated scout-bomber delivered to the US Navy in early 1939 were eighty-nine units labelled the SBC-4, with fifty units being transferred to the French Government in June 1940. All the various models of the Curtiss-Wright-designed and built dedicated scout-bombers were referred to as 'Helldivers'.

Monoplane Dive Bombers

The Northrop Company product came up with a monoplane scout-bomber designated the BT-1 in 1935. The 'T' in the letter designation code stood for Northrop, which became a division of the Douglas Aircraft Company in 1939. The BT-1 was not perfect, but the US Navy believed the design had potential and ordered fifty-four in 1936. A labour dispute delayed delivery to the US Navy until 1938. Once in service it proved unsuitable for carrier service and was pulled from use.

Fifty-seven units of an improved Northrop BT-1 design, initially referred to as the BT-2 and fitted with a more powerful engine, were delivered by Douglas to the US Navy in mid-1940. Once in production it was then labelled the SBD-1, with the letter 'D' in the designation code standing for Douglas. This first version of the aircraft was supplied to US Marine Corps squadrons. It was quickly followed by eighty-seven units of an improved SBD-2 model that same year that went to US Navy squadrons. The aircraft was nicknamed the 'Dauntless', as were follow-on versions.

In early 1941, the US Navy ordered 174 units of the SBD-3, with deliveries starting in March 1941. US entry into the Second World War in December 1941 resulted in the US Navy quickly ordering 500 additional units of the aircraft. In October 1942, the US Navy received the first of 780 units of the latest version of the Dauntless, designated the SDB-4.

Vought, not wanting to miss out on a business opportunity, wasted no time in providing the prewar US Navy with a series of new monoplane scout-bombers, beginning with fifty-four units of the SB2U-1, with the first delivered in mid-1937. It was followed by the delivery in 1938 of another model, labelled the SB2U-2, of which the US Navy ordered fifty-eight units. The final version of the aircraft ordered by the US Navy late in 1939 was the SB2U-3, and assigned the name 'Vindicator', the first of fifty-seven ordered being delivered in mid-1941.

Biplane Torpedo Bombers

The US Navy's original post-First World War dedicated torpedo bomber was the twin-engine, shore-based Martin TM-1 of which they ordered ten units. The letter 'T' stood for torpedo and the 'M' for Martin. The aircraft had its initial flight in January 1920, and was also referred to as the MTB. Again, the 'M' being for Martin and the 'TB' for torpedo bomber.

The Philadelphia Naval Yard took it upon themselves in 1922 to modify an unsuccessful Curtiss twin-engine torpedo bomber design, referred to as the R-6-L, with a more powerful engine. This resulted in a new aircraft designated the PT-1, of which fifteen were built. It was superseded the following year by an improved version, designated PT-2, of which eighteen were constructed. The letter 'P' stood for Philadelphia Naval Yard, and the 'T' for torpedo.

To come up with a more modern shore-based torpedo bomber, the US Navy in 1922 arranged for a competition between four civilian firms. The winner of the contest was a single-engine Douglas design, designated by the US Navy as the DT-2, the letter 'D' standing for Douglas and the 'T' for torpedo.

The US Navy ordered ninety-three units of the DT-2, with the building of the aircraft divided among four entities; Douglas along with two other civilian firms and the NAF, the latter building five units of an improved version in 1923, known as the DT-4.

The Douglas DT-2 and DT-4 were followed by six prototypes of the Curtiss designed CS-1 and two of the CS-2 shore-based reconnaissance plane and torpedo-bomber. However, Martin underbid Curtiss for the construction of additional planes for the US Navy. Their copies were labelled the SC-1 and SC-2, with seventy-five units built, all of which were delivered in 1925.

Douglas managed to interest the US Navy in twelve units of another shore-based torpedo-bomber, designated the T2D-1 that were delivered in 1928. The US Navy ordered eighteen additional units of another version of the T2D-1 that came with folding wings and was intended for carrier duty. That never happened, and the aircraft was confined to shore bases.

The first carrier-based torpedo-bomber for the US Navy was twenty-four units of the Martin T3M-1; deliveries started in 1926. It was followed in 1927 by 100 units of an improved version, designated the T3M-2. The Martin T3M-1 and T3M-2 were replaced onboard US Navy carriers in 1930 by another Martin aircraft designated the T4M-1, of which 102 were acquired. However, at that point, the Martin factory had been acquired by the Great Lakes Aircraft Corporation, and the aircraft designation was changed to TG-1. It was followed by an improved version in 1931 referred to as the TG-2. In total, fifty units of the TG-1 and TG-2 were built for the US Navy.

Monoplane Torpedo-Bombers

As the Great Lake TG-1 and TG-2s were beginning to show their age in 1934, the US Navy began looking for a more advanced monoplane torpedo bomber. After testing the products of three different firms, they decided a Douglas product best met their needs. That aircraft first entered service in July 1937 and was designated the TBD-1. The letters 'TB' were the new designation code for torpedo bombers and the letter 'D' obviously for Douglas. One hundred and thirty units of the TBD-1 were acquired by the US Navy between 1937 and 1939.

The TBD-1 was named the 'Devastator' in late 1941 and was assigned to all the pre-Second World War US Navy fleet carriers. In the May 1942 Battle of the Coral Sea, the Devastators performed well and helped to sink a Japanese light aircraft carrier. However, during the Battle of Midway the following month, their slow attack speed and lack of manoeuvrability made them easy targets for enemy shipboard anti-aircraft defenses, as well as defending fighters. The US Navy pulled all the remaining Devastators off its carriers shortly afterwards and confined them to secondary duties, until the last were pulled from service in 1944.

(*Opposite above*) Pictured in flight at Kitty Hawk, North Carolina, on 17 December 1903, is the Wright Brother's first powered aircraft, named the 'Wright Flyer'. That day, three flights were made by the Wright Brothers with their one-man aircraft. During the first attempt, it managed to fly 120 feet in 12 seconds. The longest flight that day was 852 feet in 59 seconds. (*US Air Force*)

(*Opposite below*) Seen here in one of a series of demonstration flights conducted at Fort Meyer, Virginia, between August and September 1908, is a Wright Brothers one-man biplane (two-wing). A young US Navy officer who was present during one of these demonstration flights wrote a report to his superiors recommending that aircraft be purchased for the service's use, but nothing came of it. (*US Air Force*)

(*Above*) Pictured is the Navy cruiser USS *Birmingham* (CL-2) fitted with a temporary sloping platform upon which the Curtiss biplane has been positioned, prior to its successful launching on 14 November 1910. Captain Chambers had originally approached the Wright Brothers to see if they would be interested in participating in the publicity stunt. However, they refused and that is when he turned to Glenn H. Curtiss for help. (*NNAM*)

(*Opposite above*) Based on what they saw with the Wright Brothers one-man biplane that was demonstrated to them at Fort Meyer, Virginia, the US Army asked them to build a larger aircraft that could support a pilot and an observer. The Wright Brothers provided the US Army what they requested in June 1909. It was named the 'Wright Military Flyer', and a museum replica is seen here. (*US Air Force*)

(*Opposite below*) Following the US Army's adoption of the world's first military aircraft, US Navy Captain Washington I. Chambers and aircraft designer and builder Glenn H. Curtiss hatched a publicity stunt in late 1910. It involved loading a Curtiss biplane onboard a Navy cruiser by crane, as seen here. The ship would then move to open water and the biplane would attempt to fly off from a temporary platform while the ship was anchored. (*NNAM*)

The civilian pilot who flew off the USS *Birmingham* (CL-2) on 14 November 1910 was Eugene Ely, seen here. A self-taught pilot, Ely's flying skill attracted the attention of Glen H. Curtiss who hired him as a demonstration pilot for his firm. Around Ely's chest are inflated bicycle inner tubes to help him float if he was forced to ditch his aircraft in the water. (*NNAM*)

Retired Navy Commander Bob Coolbaugh sits in the pilot seat of a flyable replica of the Curtiss biplane that Eugene Ely flew off the USS *Birmingham* (CL-2) on 14 November 1910. The picture was taken on the US Navy aircraft carrier the USS *George H.W. Bush* (CVN-77) on 15 November 2010, to help celebrate the centennial of naval aviation. (*DOD*)

A flyable replica of the Curtiss biplane that Eugene Ely flew off the *USS Birmingham* (CL-2) on 14 November 1910 is seen here at an airshow in 2011. The replica aircraft had been flown around the country for a number of years to celebrate the accomplishments of Glen Curtiss and Eugene Ely. (*Paul Hannah*)

To top the initial publicity stunt Captain Chambers had a temporary horizontal platform built upon the stern of the Navy cruiser USS *Pennsylvania* (ACR-4) as seen here. His plans called for Ely to both land and then launch himself off the ship in a Curtiss biplane. (*NNAM*)

(*Above*) In this picture, taken on 18 January 1911, we see demonstration pilot Eugene Ely coming in for a landing in his Curtiss biplane on the temporary horizontal platform erected on the stern of the Navy cruiser *USS Pennsylvania* (ACR-4). Ely successfully landed his biplane on the ship and took off a short time later, to the cheers of all the onlookers. (*NNAM*)

(*Opposite above*) Despite the successful demonstrations, the Secretary of the Navy informed Captain Chambers that he could not see any viable role for aircraft unless they could be brought aboard a ship and launched without the aid of an add-on platform. On 1 February 1911, Glen H. Curtiss flew his first seaplane design out to the USS *Pennsylvania* (ACR-4) and was hoisted aboard as seen here. The seaplane was then returned to the water and Curtiss took off to return to land. (*NNAM*)

(*Opposite below*) Having satisfied the Secretary of the Navy requirement for a viable seaplane, Captain Chambers ordered three aircraft in May 1911. The first to enter service was the Curtiss seaplane, designated the A-1 Triad, seen here. On the far left of the picture is Glenn H. Curtiss. To his right is Captain Chambers. (*NNAM*)

Pictured on display at San Diego, California, in February 2011, is a flyable replica of the Navy's first aircraft, a Curtiss-designed and built seaplane referred to as the A-1 Triad. In lieu of a float, the aircraft was attached to a large wooden sled. The US Navy design specifications for the A-1 Triad called for a two-man crew, either of which could fly the aircraft. *(DOD)*

Following the US Navy's acceptance of the A-1 Triad and the A-2 seaplane, Curtiss continued improving his aircraft designs. Following on his earlier successes, Curtiss came up with a more evolved seaplane which first flew in 1912, seen here, that he referred to as the F-Boat. In US Navy service it was eventually assigned the letter prefix code 'AB'. It was a single-engine biplane with a two-man crew. *(NNAM)*

The first flight of the US Navy A-1 Triad took place on 1 July 1911. Later that month, Glen H. Curtiss modified the aircraft with the addition of retractable wheels allowing the seaplane to operate on both land and water, making it an amphibian. The second Curtiss seaplane, designated the A-2, showed up in Navy service on 13 July 1911. It is shown here being lifted from the water after an accident in January 1914. (*NNAM*)

One of the US Navy Curtiss F-Boats is seen here in 1916 being launched from a catapult mounted on the bow of a US Navy cruiser. The US Navy did not employ the Curtiss F-Boats overseas during the First World War. Rather, they were confined to the training role in the United States. A late-war improved version of the F-Boat was referred to as the MF-Boat. (NNAM)

Another Curtiss-designed aircraft acquired by the US Navy during the First World War, as a pilot training platform, was the N-9 floatplane, shown here being prepared for takeoff. It was a variant of the Curtiss designed JN-4, nicknamed the 'Jenny', which he supplied to the US Army as a land-based, wheeled trainer aircraft during the First World War. (NNAM)

Curtiss also provided the US Navy during the First World War with a number of weapon-armed seaplanes intended for anti-submarine warfare duties from overseas bases. These included the twin-engine H-16 shown here, which was the final version of his H-series seaplanes that first flew in June 1914 as the H-2, and then progressed through the H-4 and H-12 versions. *(NNAM)*

Another Curtiss seaplane that served in the First World War as an anti-submarine warfare platform was the single-engine HS-2L biplane model seen here. In lieu of Curtiss-designed engines, the US Navy preferred the Liberty engine, hence the 'L' in the aircraft's designation. The HS-2L had a four-man crew. *(NNAM)*

Ordered from Curtiss by the US Navy prior to the conclusion of the First World War were four, large, three-engine seaplanes intended as anti-submarine warfare (ASW) platforms, labelled NC-1 through NC-4, with NC-4 seen here. None were finished before the conclusion of the conflict, but were involved in some postwar publicity flights. (*NNAM*)

Following the First World War, the first observation floatplane intended to be catapulted off US Navy ships was the Vought VE-9H biplane shown here. Only four were built in this version, with all delivered in 1927. It was unarmed and had a crew of two. The aircraft was based on the Vought VE-7 single-seat fighter first adopted by the US Navy in 1920. (*NNAM*)

The follow-on to the Vought VE-9H catapult-launched observation floatplane was the Vought O3U-3 biplane, seen here, that began appearing in US Navy service in 1934. It had a two-man crew and was powered by a single 600 hp engine. Self-defense armament consisted of three machine guns. (*NNAM*)

Pictured on the *USS West Virginia* is a catapult-launched observation floatplane known as the Curtiss SOC-1 Seagull. The prototype of the aircraft flew in March 1934 and met all its performance specifications. Deliveries of the SOC-1 Seagull began the next year, with 135 units ordered. (*NNAM*)

(*Above*) The Vought OS2U Kingfisher, three of which are seen here at a US Navy training base, had a maximum speed of 176 mph. It was primarily employed during the Second World War as a naval gunfire observation plane. An improved model, with armour and self-sealing fuel tanks, was known as the OS2U-2 Kingfisher. (*NNAM*)

(*Opposite above*) In addition to the floatplane version of the Curtiss SOC-1 Seagull, there was another model ordered by the US Navy in late 1936 that was wheeled, as shown here. It was assigned the designation SOC-2 Seagull and the US Navy took forty into service as general utility aircraft. A small number of the SOC-2 Seagulls were later modified by the US Navy with arresting gear to land on carriers. With this modification the aircraft became the SOC-2A. (*NNAM*)

(*Opposite below*) Intended as a replacement for the Curtiss SOC-1 Seagull, the US Navy took into service, beginning in 1940, the Vought OS2U Kingfisher. Seen here is the OS2U-1 version. Unlike its predecessors that were all biplanes, the OS2U Kingfisher was a monoplane. Like the Curtiss SOC-1 Seagull, it had a two-man crew and could also be converted into a landplane. (*NNAM*)

HOISTING AERO ON U.S.S. OKLA. A

(*Above*) Prior to the development of specialized catapult-launched observation floatplanes for the US Navy, trials were conducted by the service right after the First World War to see if wheeled aircraft could be launched from short launching platforms mounted on battleship main gun turrets. Pictured, a French-designed and built Nieuport 28 fighter ready to be brought aboard the *USS Oklahoma*. (*NNAM*)

(*Opposite above*) Pictured is one of the experimental launching platforms mounted on a main gun turret of a US Navy battleship, shortly after the First World War. Visible is a French *Hanriot* biplane on the rear of the launching platform, which consisted of a supporting metal framework that bolted to the barrels of a battleship's main gun turret which was covered with wooden planking. Once launched, the aircraft had no way to be recovered aboard ship. (*NNAM*)

(*Opposite below*) Shown is a twin-engine US Navy PD-1 flying boat sometime in the 1920s. It was an improved copy of the British-designed First World War-era F-5-L flying boat. The F-5-L flying boat itself was an improved version of the British-designed *Felixstowe* F.5, which in turn was an improved version of the Curtiss H-12 supplied to the British during the First World War. Besides Douglas, three other companies built versions of the aircraft for the US Navy. (*NNAM*)

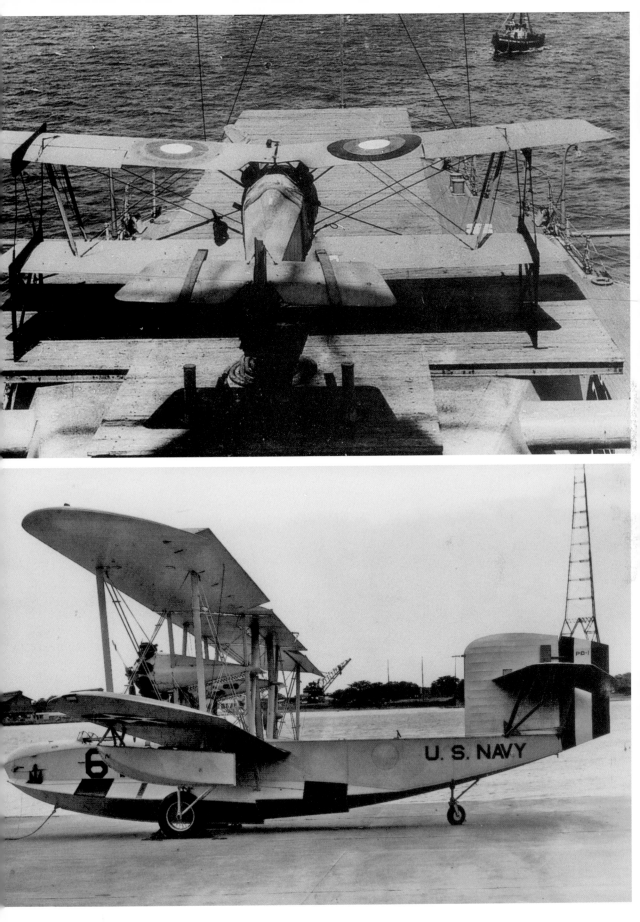

The Martin P3M flying boat shown here was actually a Consolidated Aircraft Company design, however, Martin had won the production contract from the US Navy by offering them a lower bid to build the aircraft. Consolidated refused to supply their competitor with the aircraft blueprints and Martin therefore had to reverse-engineer the prototype to build their version of the aircraft for the US Navy. *(NNAM)*

On the ground in 1940 is a twin-engine Consolidated P2Y flying boat. The original prototype had been designed with three engines, but it was soon decided that it was not needed, and the third engine was deleted from the production units delivered to the US Navy, beginning in 1931. In total, seventy-eight of the Consolidated P2Y flying boats were built. It eventually appeared in three different models, the P2Y-1, P2Y-2, and the P2Y-3. *(NNAM)*

The replacement for both the Martin P3M and the Consolidated P2Y flying boats was another design from the latter firm that was known as the PBY Catalina, seen here in flight. The twin-engine aircraft offered both increased range and load-carrying ability compared to its direct predecessors. The US Navy was so enthusiastic about this aircraft's capabilities that the initial version, the PBY-1 was ordered into production in 1935. (*NNAM*)

The PBY-1 through PBY-3 versions were strictly flying boats. When the follow-on models were fitted with retracting wheeled landing gear, they became amphibians. Reflecting this new feature, follow-on variants of the aircraft had the letter 'A' added to their designations, resulting in the PBY-5A and PBY-6A, a beautiful restored example of the latter is shown here. (*Loren Hannah*)

(*Above*) Thinking about a second twin-engine flying boat to supplement the Consolidated PBY Catalina, the US Navy took into service, in September 1940, the first of numerous versions of the Martin PBM Mariner. The PBM Mariner was an amphibian with retractable wheeled landing gear. Sailors are shown washing off the corrosive salt water from a PBM Mariner just back from a patrol. (*RWP*)

(*Opposite above*) In total, Martin built 1,366 units of the PBM Mariner for the US Navy. Here we see a group of sailors trying to push the aircraft out into the water for takeoff. The PBM Mariner was armed with up to eight machine guns. It could carry 4,000 lbs of ordnance, consisting of conventional bombs and depth charges. (*RWP*)

(*Opposite below*) The US Navy's interest in airships, also referred to at the time as lighter-than-air ships, began on 20 April 1915 with the purchase of an American-designed and built non-rigid model airship. Non-rigid airships are referred to as 'blimps' in today's lexicon. Eventually sixteen non-rigid airships were ordered by the US Navy for use during the First World War, an example is seen here. None were deployed overseas during the conflict. (*NNAM*)

(*Above*) Hanging from the ceiling of an American museum is the control cab of K-47, a Goodyear-designed and built K-class non-rigid airship employed by the US Navy during the Second World War. The K-class non-rigid airships survived in US Navy service until 1959. The control cab on these non-rigid airships was forty feet long. (*Paul Hannah*)

(*Opposite above*) Following the First World War, the US Navy continued to experiment with non-rigid airships. Eventually, the first of the Goodyear K-class non-rigid airship series entered service with the US Navy in 1938. An example is seen here in April 1949. During the Second World War, over 134 K-class non-rigid airships were built as antisubmarine warfare (ASW) platforms. (*RWP*)

(*Opposite below*) In the 1920s, the US Navy decided that they needed to have airships capable of long-range reconnaissance missions. As non-rigid airships were not up to the task, the US Navy experimented with large rigid airships, taking five into service between 1923 and 1933. Pictured is the last rigid airship commissioned by the US Navy, the USS *Macon* (ZRS-5). (*NNAM*)

(*Above*) The interwar US Navy rigid airships, the USS *Akron* (ZRS-4) and the USS *Macon* (ZRS-5) were designed with an internal hangar bay that had enough room to house four Curtiss F9C Sparrow Hawk biplane fighters. The planes were launched and recovered with the assistance of a retractable trapeze-like device seen here. (*NNAM*)

(*Opposite above*) Originally designed during the First World War as a two-seat trainer for the US Army, the Vought VE-7 Bluebird so impressed the US Navy with its operational capabilities that a version of it was adopted as the service's first fighter in 1920. An improved version seen here was designated the VE-9 and appeared in service in 1922. (*NNAM*)

(*Opposite below*) The US Navy's first dedicated carrier fighter, the Boeing FB-5 is seen here. Based on the design of a late-production First World War German fighter, the one-man FB-5 entered service with the US Navy in 1926. Earlier versions, the FB-1 through FB-4 were basically test models that evolved into the final FB-5 variant. The aircraft had a maximum speed of 159 mph and was armed with two machine guns. (*NNAM*)

The US Navy took into service the Boeing F4B-4 fighter, shown here as a museum display, in the early 1930s. It was the replacement for the Boeing FB-5 fighter. Earlier versions of the plane were labelled F4B-1 through F4B-3. A total of ninety-two units of the F4B-4 were acquired by the US Navy, who transferred twenty-one units to the US Marine Corps. (*Christopher Vallier*)

A young sailor poses for the photographer in front of a line of Curtiss F6C-1 Hawk fighters. The aircraft entered US Navy service in the late 1920s, and came in a number of different models, beginning with the F6C-1 and ending with the F6C-4. Some went on to serve aboard carriers, with others being transferred to the US Marine Corps. (*NNAM*)

In the early 1930s, Curtiss-Wright supplied the US Navy with a new carrier fighter, designated the F8C Falcon series. Eventually they were passed on to the US Marine Corps who employed them as observation planes with a backup role as dive bombers. The final production version of the F8C series is seen here in US Marine Corps markings and was referred to as the OSC-1 Helldiver. (NNAM)

The Grumman FF-1 fighter seen here entered US Navy service in 1933. Powered by a 750 hp engine, the two-man aircraft could reach a maximum speed of 201 mph. A modified version with a different engine was designated the SF-1 and entered service in 1934. The FF-2 was a modified version of the FF-1 with dual controls and employed as a trainer. (NNAM)

(*Above*) Problems with the Grumman FF-2 flight characteristics led to the US Navy asking the firm to come up with an improved model that entered service in 1936 as the F3F-1 fighter, of which fifty-four were built. The next model introduced in 1937 was the F3F-2, an example of which is seen here on the flight deck of a carrier in a less-than-perfect landing. (*NNAM*)

(*Opposite above*) The Brewster F2A-2 fighter seen here was named the Buffalo, and was a mid-wing monoplane intended for carrier service. Initially ordered by the US Navy in June 1938, the first model was designated the F2A-1 and began showing up in service the following year. The third and final version of the aircraft was referred to as the F2A-3. (*RWP*)

(*Opposite below*) Preparing for launching off a US Navy carrier's flight deck in May 1942 is a squadron of Grumman F4F-3 Wildcats. The Wildcat was armed with four .50 caliber machine guns. The next version of the aircraft, designated the F4F-4, was armed with six .50 caliber machine guns. Both versions of the Wildcat were flown by the US Marine Corps during the Second World War. (*NNAM*)

At a Grumman assembly facility in December 1941 is an array of wingless F4F-3 fuselages. The first order for the F4F-3 was not from the US Navy, but from the French government in early 1939. However, before they could be delivered, the German military overran the country, in the spring of 1940. The US Government then had the French order diverted to the Royal Navy. (*NNAM*)

Within the hangar deck of the US Navy aircraft carrier USS *Enterprise* (CV-6) on 28 October 1941 is a crew of mechanics working on a Grumman F4F-3 Wildcat. The aircraft was a mid-wing monoplane. The F4F-3 had a length of 28 feet 9 inches and a wingspan of 38 feet. (*RWP*)

A less-than-perfect landing of a US Navy Grumman F4F-4 Wildcat has resulted in the collapse of its landing gear. Despite being inferior on paper to the Japanese A6M Zero in some operational parameters, American Wildcat pilots outflew their opponents during the Second World War and obtained a kill-to-loss ratio of 6.9 to 1. *(National Archives)*

Grumman stopped building the F4F-4 Wildcat in early 1943. The US Navy assigned the continued production of the aircraft to the Eastern Aircraft Division of General Motors (GM). The original GM production version of the F4F-4 was labelled the FM-1 and was a near identical copy of the Grumman-built plane, except it only had four .50 caliber machine guns. The follow-on version, seen here at an airshow, was designated the FM-2. *(Christopher Vallier)*

Pictured is another restored example of an FM-2 Wildcat at an airshow. The Eastern Aircraft Division of General Motors built 5,280 units of the FM-1 and FM-2 during the Second World War, of which 503 were the FM-1 model and the remaining 4,777 units the FM-2 model. In total, Grumman built 1,971 units of the F4F-3 and F4F-4. *(Paul Hannah)*

The Martin BM-2 dive bomber shown here was powered by a 625 hp engine that gave it a maximum speed of 146 mph. It was armed with two 7.62 mm machine guns for self-defense. The aircraft had a wing span of 41 feet and was 28 feet 9 inches in length. *(NNAM)*

In flight is a formation of US Navy Curtiss SBC-4 Helldivers, the last prewar biplane dive bomber. The prefix code letters 'SB' stood for scout-bomber, the new designation for dive bombers, which replaced the previous prefix code letters 'BF' that had stood for bomber-fighter. The SBC-4 Helldiver was the first US Navy biplane scout-bomber to have retractable landing gear. (*NNAM*)

A US Navy Curtiss F11C-2 Hawk is shown in this picture. Reflecting its secondary job as a dive bomber, it later received the designation BF-2C Hawk, the letter prefix code 'BF' standing for bomber-fighter. An improved version of the BF-2C Hawk was labelled the BF2C-1 Goshawk. The BF-2C Hawk was 22 feet long and had a wingspan of 31 feet 6 inches. (NNAM)

The monoplane scout-bomber that set the general pattern for all those that followed it into US Navy service, was the Northrop BT-1, seen here with a temporary experimental tricycle landing gear. Fifty-four where ordered by the US Navy in 1936, with the first delivered for carrier use in 1939. The plane was not a success in use due to some unresolved design issues that made it difficult to fly and land on carrier flight decks. (NNAM)

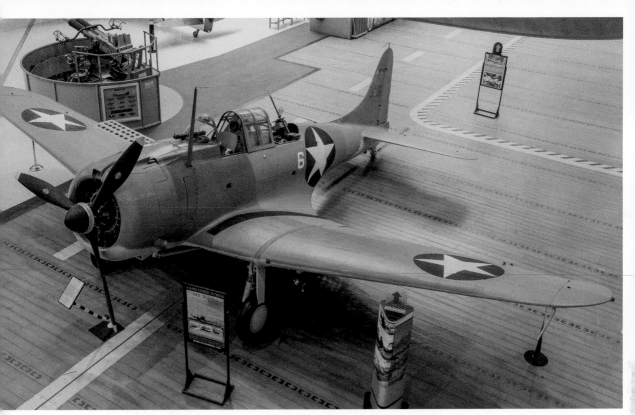

In April 1938, the US Navy tested an improved version of the Northrop BT-1 scout-bomber, which they ordered into production in 1939. By this time Northrop had been acquired by Douglas and the new scout-bomber series was designated the SBD Dauntless, with models ranging from one through six, with the last three versions being built during the Second World War. Pictured on display at an American museum is an SBD-3. *(Paul Hannah)*

Shown is a US Navy Douglas SBD-3 Dauntless scout-bomber on the flight deck of the *USS Hornet* (CV-8) during the Battle of Midway in June 1942. In the course of that famous engagement, US Navy Dauntless scout-bombers accounted for four Japanese aircraft carriers and a heavy cruiser. The SBD series served in frontline US Navy carrier service until the middle of 1944. *(RWP)*

(*Above*) A US Navy Douglas DT-2 aircraft is shown launching a 1,800 lb torpedo in 1925, during a training exercise, with the letter 'D' standing for Douglas and the 'T' for torpedo. The DT-2 could be fitted with wheels or twin floats. It had a length of 37 feet 8 inches and a wingspan of 50 feet. The Navy ordered sixty-four units of the DT-2. (*NNAM*)

(*Opposite above*) On display is the only surviving example of the 260 units of the Vought SB2U series of scout-bombers, assigned the name the 'Vindicator'. This particular aircraft is an SB2U-2. Vought also supplied the US Navy with an SB2U-1 and SB2U-3 model of the aircraft. The SB2U series were in US Navy frontline carrier service from December 1937 until late 1942. (*Paul Hannah*)

(*Opposite below*) With the introduction of carriers to the US Navy, a need arose for a biplane torpedo bomber that could be launched and recovered from the new ships. The answer was the three-man Martin T3M-1, which the US Navy first took into service in 1926. It was superseded the following year by the improved T3M-2, seen here flying in formation, that could be configured with wheels or floats. (*NNAM*)

(*Above*) The replacement for the Martin T4M-1 torpedo bomber and its Great Lake Company-built units which exited US Navy frontline carrier service in 1938, was the Douglas TBD-1 Devastator, seen here. It was the first carrier-capable US Navy monoplane and began entering service in 1937. It had a 900 hp engine that gave it a maximum speed of 206 mph. (*NNAM*)

(*Opposite above*) A third model of the Martin T3M series of torpedo bombers was provided with a more powerful engine and labelled the T4M-1 by the US Navy, an example of which is seen here. When the Great Lakes Company took over the Curtiss plant that built the T4M-1 in late 1928, the aircraft was re-labelled the TG-1, and another model with a more powerful engine became the TG-2. (*NNAM*)

(*Opposite below*) In the 1930s, the US Navy took into service a number of two-seat training aircraft, including this restored Stearman Model 75 seen here. In US Navy service it was assigned the designation code 'NS', with the 'N' standing for trainer and the 'S' for Stearman. The firm became a subsidiary of Boeing in 1939. (*Christopher Vallier*)

The restored North American SNJ-4 Texan seen here was initially ordered by the US Navy in 1936. There were a number of different models of the Texan taken into service by the US Navy as trainers. These were labelled SNJ-1 through SNJ-6. The letters 'SN' in the designation prefix code stood for scout-trainer. *(Christopher Vallier)*

First ordered in November 1940 by the US Navy was the two-seat Curtiss-Wright SNC-1 Falcon, seen here. It was originally designed as a light fighter but was only employed as a training aircraft prewar and during the Second World War. In total the US Navy would take 305 units of the Falcon into service by 1945. *(NNAM)*

In the early 1930s, the US Navy ordered from Grumman a new biplane utility amphibian aircraft designated the JF-1, the letter 'J' standing for utility. Progressively improved models were all eventually referred to as the 'Duck'. The restored model pictured is the J2F-4. The first of the J2F series Ducks appeared in service in 1936. *(Christopher Vallier)*

At some point in the late 1920s or early 1930s, the US Navy acquired the Ford 5-AT-C civilian passenger plane named the 'Trimotor'. In US Navy service it was designated the JR-3. The example of the Trimotor seen here is in US Army Air Corps markings. Ford built 199 of the Trimotors in five different versions between 1926 and 1933. *(US Air Force)*

Chapter Two

Second World War Aircraft

Because of the sometimes long lead time between development of an aircraft and its successful acceptance into service, the US Navy has always been forced to begin thinking about the next generation of aircraft as soon as possible. Reflecting this need, the US Navy contracted with Grumman in June 1941 to begin work on a replacement for their Wildcat fighter only a few months after the first model of that plane began showing up on US Navy carriers.

The Grumman replacement for the Wildcat was the F6F-3 Hellcat. It was for all intents a much improved and larger Wildcat. Both aircraft had the short and stubby appearance that was the hallmark of the Grumman pre-Second World War fighters. The Hellcat design benefited from user input provided both by American pilots, as well as foreign users of the Wildcat who had seen combat against the Axis overseas.

The Hellcat first entered into US Navy service in January 1943, but did not see action until August that year. As the tempo of Hellcat production increased, the plane soon displaced the Wildcat with both US Navy and US Marine Corps squadrons on US Navy fleet carriers, but it was too big for the escort carriers where the Wildcat remained. The Hellcat was armed with six .50 caliber machine guns. Once engaged in combat the plane quickly demonstrated its superiority over existing Japanese fighters.

Grumman did not stop trying to improve the original F6F-3 Hellcat and a second model, designated the F6F-5, began appearing on US Navy carriers in the summer of 1944. Both types of Hellcat could carry either bombs or rockets, the latter being a late-war addition equally useful against seaborne or land targets. In total, over 12,000 units of the Hellcat series were built by the time production ended in 1945. It also saw wartime service with the Royal Navy, and post-war with the French and Uruguayan Navies.

The Bent Wing Fighter

A plane whose genesis began with a US Navy contract awarded in June 1938 for the building of a single prototype fighter eventually evolved into the famous Vought F4U Corsair. The first prototype of the aircraft flew in May 1940, with its trademark inverted gull wings, and reflected the company's designers bringing together a stream-lined airframe with the most powerful engine then available.

During a demonstration flight in October 1940, the prototype Corsair attained the unheard-of top speed of 404 mph for a single-engine aircraft in level flight. This did not result in the quick awarding of a production contract, as overseas combat reports indicated that any new fighter being considered had to be both armored and up-gunned, as well as having self-sealing fuel tanks. This forced a redesign of the plane that pushed back the awarding of a production contract to Vought for the Corsair until June 1941.

The first production Corsair flew in June 1942 and was designated the F4U-1. However, early problems adapting it to US Navy carrier use resulted in the aircraft being confined to operating from shore bases by the US Marine Corps. Those US Navy squadrons initially equipped with the F4U-1 and intended to serve on carriers were re-equipped with Hellcats prior to being sent into combat. Despite the aircraft initially being restricted to land bases, it racked up an impressive kill rate over its Japanese counterparts, second only to the carrier-based Hellcat.

Following the F4U-1 model, Vought also came up with other versions of the Corsair. These included the F4U-1A (not an official US Navy designation), the F4U-1C, and the F4U-1D. The 'C' model of the Corsair was armed with four 20 mm automatic cannons for ground attack, two in each wing, rather than the six .50 caliber machine guns, three in either wing, fitted to the original F4U-1 and the improved F4U-1A model. With the 'D' model of the Corsair, the aircraft reverted to six .50 caliber machine guns.

The introduction of the F4U-1D model into US Navy service marked the addition of under-fuselage and under-wing attachment points to the aircraft, known as 'hard points' (a term that will be used hereafter in the text). These hard points could carry either ordnance (bombs and rockets), or extra fuel tanks, and were eventually fitted to previous models of the Corsair.

A night-fighter variant of the F4U-D1, designated the F4U-2 equipped with only five machine guns and a radar in its starboard wing, was first delivered to the US Navy in January 1943. It was the night fighter models of the Corsair that were the first to be approved for carrier use in early 1944 by the US Navy.

The final version of the Corsair series to see combat in the Second World War (in very small numbers) was designated the F4U-4. It was armed with six .50 caliber machine guns. Like its predecessors it came in a number of variants including the F4U-4C armed with four 20 mm automatic cannon. There was also an F4U-4N night fighter, and an F4U-4P photo-reconnaissance model.

Due to high production demands for the Corsair series, the US Navy brought in both Goodyear and Brewster to build near-identical copies. The Goodyear-built version of the Vought F4U-1A was referred to as the FG-1A and the Goodyear-built version of the Vought F4U-1D was labelled the FG-1D.

There had been plans for Goodyear to build a copy of the Vought F4U-4 as the FG-4, but the end of the Second World War resulted in the cancellation of that production run. The Brewster built copy of the Vought F4U-1 was labelled the F3-A1, but never saw front-line service due to poor manufacturing and quality control.

Carrier-Based Daytime Corsairs

Due to a shortage of US Navy Hellcat fighter squadrons in late 1942, the senior leadership of the US Navy decided to rescind their ban on flying the Corsair fighter series off carriers during daylight hours. By this time, a number of design fixes had been applied to the aircraft to increase its suitability for carrier operations.

The first Corsair equipped fighter squadrons to be assigned to a US Navy carrier were not US Navy but US Marine Corps—VMF-124 was assigned to the USS *Essex* (CV-9) on 28 December 1944. Eventually, US Navy Corsair fighter squadrons were assigned to carriers in the last year of the war. However, the majority of US Navy fighter squadrons flew the Hellcat in 1945.

Postwar Production

By the time the Second World War had ended, American factories had built approximately 12,000 units of the Corsair in a variety of models. The usefulness of the Corsair ensured that production continued into the early postwar years for the US Navy, where it remained a frontline carrier close-air support aircraft through the Korean War. In total, 750 units of the Corsair series would be built postwar, with production finally concluded in 1953.

The first postwar built model of the Corsair series was the F4U-5, with a night fighter version designated the F4U-5N. Another slower, but more heavily armored version of the aircraft, intended strictly for the ground attack role for the US Marine Corps, was referred to as the AU-1. All three saw action during the Korean War and accounted for a small number of enemy prop-driven planes, as well as a single jet-powered enemy fighter.

Too Late for Combat

As already mentioned, the US Navy always tried to look as far into the future as possible to figure out what types of carrier aircraft it would need. Reflecting this train of thought, two additional fighters went into production at the end of the Second World War. These included the single-engine Grumman F8F-1 Bearcat, optimized for the interceptor role, and the twin-engine Grumman F7F-1 Tigercat.

The Bearcat was intended as the replacement for the Hellcat on US Navy carriers, with the first delivery of the aircraft taking place in February 1945. However, squadrons equipped with the Bearcat would not see combat with the US Navy or US Marine Corps during the Second World War.

Despite the impressive performance of the Bearcat, prop-driven fighters were quickly being replaced by jet-powered aircraft in the immediate postwar era, and it was pulled from both the US Navy and US Marine Corps frontline inventory before the Korean War. A total of 1,265 units of the Bearcat were constructed out of an original order in October 1944 for 2,023 units, the rest having been cancelled.

The first production units of the F7F-1 Tigercat were handed over to the US Navy in April 1944. A two-seat night-fighting model of the Tigercat was designated as the F7F-2N and was delivered in October 1944. It was followed by the delivery of another single-seat model, labelled the F7F-3, beginning in March 1945.

Testing conducted by the US Navy in April 1945 showed that the Tigercat was unsuitable for carrier use due to a number of design issues. The US Navy therefore transferred all of them to the US Marine Corps for the land-based close-air support role, however, it showed up in squadron service too late to see combat during the Second World War. The aircraft saw action during the early part of the Korean War with the US Marine Corps. The enemy introduction of jet-powered fighters resulted in the Tigercat being withdrawn from combat use in Korea in 1952. It lasted in US Marine Corps service until 1954. A total of 364 units of the Tigercat series were built.

The Last Scout-Bombers

Ordered by the US Navy in November 1940 from Curtiss-Wright was a monoplane scout-bomber, designated the SB2C-1, which the company named 'Helldiver'. It was intended to replace their earlier Curtiss SBC-3 and SBC-4 scout-bombers, both being biplanes, also named Helldivers. The SB2C-1 was also ordered by the US Army Air Forces, who had been very impressed by the German Air Force employment of dive bombers during their attack on Poland in September 1939.

The SB2C-1 Helldiver did not enter into production until mid-1942 due a number of design problems and production bottlenecks. Only 200 units of the SB2C-1 were built for the US Navy and none saw combat. An improved version, labelled the SB2C-1C, first saw combat with the US Navy in late 1943, with a total of 778 units built. The SB2C-3 variant had a more powerful engine fitted and 1,112 units were constructed. Fitted with a radar, the SB2C-3 was designated the SB2C-3E. The letter 'E' in the aircraft's designation stood for electronics.

Also assembled were 2,045 units of a modified version of the SB2C-1, known as the SB2C-4. It was fitted with underwing hard points for air-to-surface rockets or bombs. Equipped with a radar unit the plane became the SB2C-4E. An upgraded version of the SB2C-4 without a radar was designated the SB2C-5; a total of 970 units of the SB2C-5 were built, with another 2,500 units cancelled due to the end of the Second World War. In total, Curtiss-Wright built 5,516 units of the SB2C series.

Two Canadian firms built copies of the Curtiss-Wright SB2C-1, SB2C-3, and SB2C-4E Helldivers. These included Fairchild-Canada and Canadian Car and Foundry.

The 300 copies of the SB2C-1, SB2C-3, and SB2C-4E constructed by Fairchild-Canada were respectively labelled the SBF-1, SBF-3, and SBF-4E. The 860 copies built by Canadian Car and Foundry were respectively labelled the SBW-1, SBW-1B, SBW-3, and SBW-4E. Canadian Car and Foundry also built a version of the SB2C-5 that was referred to as the SBW-5.

Wartime Impressions

Combat service of the SB2C series of Helldivers with the US Navy was extremely mixed, with many feeling that it was a badly-designed aircraft that reflected poorly on its designer and builder. The issues, both real and perceived, that bedeviled the SB2C series badly sullied the reputation of Curtiss-Wright and contributed to it being the last aircraft acquired by the US Navy from the firm.

Most of those in the US Navy who had to deal with the SB2C series during the Second World War felt that its predecessor, the Douglas SBD Dauntless, was a better aircraft, despite the Curtiss-Wright SB2-C being superior in its operational parameters, except in range. A total of thirty US Navy and twenty-five US Marine Corps squadrons flew the aircraft during the Second World War. The US Marine Corps inventory of SB2C series aircraft were primarily land-based.

Another scout-bomber that was approved for production by the US Navy prior to Pearl Harbor, but which did not begin coming off the factory floor until 1943, was the Brewster SB2A Buccaneer. It was the Brewster replacement for their earlier SBN-1. It was not a success in the scout-bomber role and was quickly transferred to training duties. The Royal Navy, which received the SB2A under Lend-Lease, also concluded that it was unfit for combat, and confined it to secondary duties. Brewster built a total of 771 units of the SB2A.

The US Navy's Last Torpedo-Bomber

The US Navy's wartime-built torpedo bomber was the Grumman TBF-1 Avenger. It was the intended replacement for the pre-war designed Douglas TBD-1 Devastator. The origins of the TBF-1 began in early 1940, when the US Navy asked both Vought and Grumman for a state-of-the-art torpedo bomber. Vought won the competition, but their plane, designated the TBY Sea Wolf by the US Navy, was besieged by a host of design and production problems that resulted in it showing up in service late in the Second World War, and never seeing frontline combat service.

By default, the TBF-1 torpedo bomber filled in for the TBY during the conflict, with the first delivery taking place in January 1942. A follow-on model, designated the TBF-1C, had larger fuel tanks to increase the aircraft range, and two wing-mounted, forward-firing .50 caliber machine guns, in place of the single forward-firing fuselage-mounted .30 caliber in the original model of the aircraft. The aircraft was powered by a 1,700 hp engine.

There were additional versions of the TBF-1 converted for roles other than torpedo-bomber, with a letter or letters added to the end of their aircraft designation to identify their new jobs. These new assignments included radar equipped units labelled the TBF-1D and TBF-1CD. Those modified for photo-reconnaissance work were referred to as TBF-1P and TBF-1CP. A model specially equipped for poor weather conditions was designated the TBF-1J. In total, Grumman built 1,526 units of the TBF-1 series, of which 465 units went to America's wartime Allies.

Due to the high wartime demand for the TBF series, it was decided to have General Motors build copies of the aircraft. The US Navy designated the General Motors copy of the Grumman TBF-1 model as the TBM-1 and the Grumman TBF-1C model as the TBM-1C. These aircraft started coming off the assembly line in September 1942. In total, General Motors constructed 7,546 units of the Grumman designed TBF-1 and TBF-1C, which, like the Grumman-built product, was also modified for other roles and assigned additional letter designations to define their purpose.

The last model of the Grumman designed TBF series of torpedo-bombers was a General Motors development of the aircraft, with a more powerful 1,900 hp engine and strengthened wings to carry more ordnance, and electronics, such as radar units. It was designated the TBM-3 and the initial delivery of the model began in April 1944. Of the 4,011 units of the TBM-3 built, many served in a variety of roles, like earlier models of the aircraft, and were labelled with additional letter designations. Two hundred and twenty-two units went to the Royal Navy.

Scout-Bomber and Torpedo-Bomber Demise

With the introduction of the F6F Hellcat and the F4U Corsair into service, the US Navy now had carrier fighters that were capable of hauling into combat an ordnance load that almost equalled what the existing scout-bombers and torpedo-bombers could tote. Unlike the bombers that were vulnerable to enemy fighters both before and after delivering their ordnance, the new fighters could defend themselves. This was another reason for the growing numbers of fighters on late-war US Navy carriers taking the place of scout-bombers and torpedo-bombers.

Seaplane Patrol Bombers

Besides the very capable twin-engine pre-Second World War Consolidated PBY series and the Martin PBM series, built in large numbers during the Second World War, the US Navy ordered from Consolidated in 1939 six test units of a four-engine seaplane designated the PB2Y-2. Testing went well and an improved production version labelled the PB2Y-3 began showing up in US Navy service in early 1942. Eventually, Martin built 210 units of the PBY-3 Coronado series. Most were upgraded in wartime to the PBY-5 standard.

The Martin Company came up with a very large four-engine seaplane ordered by the US Navy in August 1938. However, the first flight of the prototype, designated the XPB2M-1 Mars, did not take place until July 1942. By that time, the US Navy realized the aircraft's design was not up to wartime standards as a patrol bomber. However, in January 1945, it ordered twenty transport versions of the plane, labelled the JRM Mars, with the first being delivered in June 1945. With the war ending in September 1945, the order was cut back to only five units.

Land-Based Patrol Bombers

Due to the high demand for maritime patrol bombers during the Second World War, the US Navy took into service a number of land-based multi-engine aircraft. This upended the 1931 agreement between the US Army Air Corps and the US Navy, restricting the latter to employing only seaplanes and ship-based aircraft.

A twin-engine aircraft originally designed as a pre-war passenger plane by Lockheed and modified into a maritime patrol bomber was designated the PBO-1. Twenty were acquired by the US Navy and a squadron based in Newfoundland sank a single German submarine in March 1942.

The PBO-1 was followed into service by 1,600 units of a similar model, also built by Lockheed, designated the PV-1 Ventura. A large number of the units of the PV-1 went to America's wartime allies. A follow-on model for the US Navy was referred to as the PV-2 Harpoon, with 535 units built, with delivery beginning in March 1944. Most saw service in the Pacific Theater of Operation (PTO) during the Second World War.

For long-range maritime patrol bomber duties the US Navy decided that land-based four-engine aircraft were superior to four-engine seaplanes. This came about due to the successful use of the Consolidated four-engine B-24 Liberator heavy bomber in that role by the RAF Coastal Command. The US Navy's request for an allotment of modified Liberators was approved in July 1942, and the aircraft was labelled the PB4Y-1.

A few PB4Y-1s were later configured for the long-range reconnaissance role and received the designation PB4Y-1P. Approximately 1,000 units of the PB4Y-1 were taken into US Navy service during the Second World War. Its usefulness resulted in it being employed in the immediate postwar era.

In May 1943, the US Navy ordered from Consolidated 739 units of a version of the Liberator designed specifically for naval use, which they designated the PB4-Y Privateer. However, very few made it into service before the Second World War ended. The aircraft remained in US Navy service until 1954. In March 1943, Consolidated had merged with the Vultee Aircraft Company and eventually became known as Convair.

Miscellaneous Aircraft

Other aircraft originally intended for use by the US Army Air Forces during the Second World War were acquired by the US Navy during the conflict to perform different jobs. These included various models of the twin-engine North American B-25 Mitchell medium bomber, of which initial delivery began in January 1943.

The US Navy assigned its inventory of 706 B-25 Mitchell medium bombers the designation code PBJ-1, and kept the US Army Air Forces model letter prefixes, as the planes were not modified in any way. Rather than being employed as maritime patrol bombers, they were all transferred to the US Marine Corps. They employed them as rocket-equipped attack aircraft in the PTO, generally going after Japanese shipping, beginning in March 1944.

Another US Army Air Forces medium bomber adopted by the US Navy during the Second World War was an early version of the Martin B-26 Marauder that had been modified in 1943 as an unarmed target tug, and eventually designated the TB-26B. In US Navy service it became the JM-1. Another unarmed target tug version of a later production B-26 was labelled the TB-26G. It became the JM-2 in US Navy service. Both aircraft were employed as target tugs and utility planes by the US Navy, with a total of 272 units being taken into the inventory.

Also picked up for service by the US Navy during the Second World War were several pre-war designed multi-engine aircraft originally designed for the civilian passenger plane industry, the best known being the militarized version of the twin-engine Douglas DC-3. It was designated the C-47 Skytrain by the US Army Air Forces and in US Navy service the aircraft was referred to as the R4D.

Another militarized prewar designed twin-engine passenger plane acquired by the US Navy was the Curtiss C-46A Commando, designated as the R5C1, all of which went to the US Marine Corps. A four-engine pre-war-designed passenger plane acquired by the US Navy from the US Army Air Forces during the Second World War was the Douglas C-54 Skymaster. In US Navy service it was labelled the R5D series and a total of 183 units of various versions were taken into the inventory.

Helicopters

The helicopter, which appeared at the tail end of the Second World War, was of little interest to the US Navy at that time. Rather than be bothered with it, they assigned helicopter development to the US Coast Guard, which was attached to the US Navy during the Second World War. Sikorsky provided the US Coast Guard several helicopter models, designated by the US Navy as the HNS-1, the HOS-1, and the HO2S-1. It was only after the Second World War, as helicopter technology progressed, that the US Navy began to see the usefulness of this new flying machine and took over its development.

(*Above*) The first production unit of the Grumman F6F Hellcat was delivered to the US Navy in December 1942 and designated the F6F-3. An example is seen here on the athwartship hangar catapult of a carrier during the Second World War. External features that identify early production F6F-3s include the forward canted antenna mounted behind the cockpit, and the wing-mounted machine gun fairings. (*NNAM*)

(*Opposite above*) In this March 1944 photograph we see a Grumman F6F-3 Hellcat that has come to grief during its recovery on the flight deck of a carrier after suffering engine problems – note the oil streaks on the cowling. During the Second World War, US Navy and US Marine Corps Hellcats accounted for 5,155 Japanese aircraft, more than half of the 9,258 Japanese planes shot down during the conflict. (*NNAM*)

(*Opposite below*) A US Navy Grumman F6F-3 Hellcat is shown fitted with a large external fuel tank that could be jettisoned if needed. This particular aircraft can be identified as a later production example of the F6F-3 by the now vertical oriented antenna mounted behind the cockpit. A total of 4,646 units of the F6F-3 were built by Grumman. (*NNAM*)

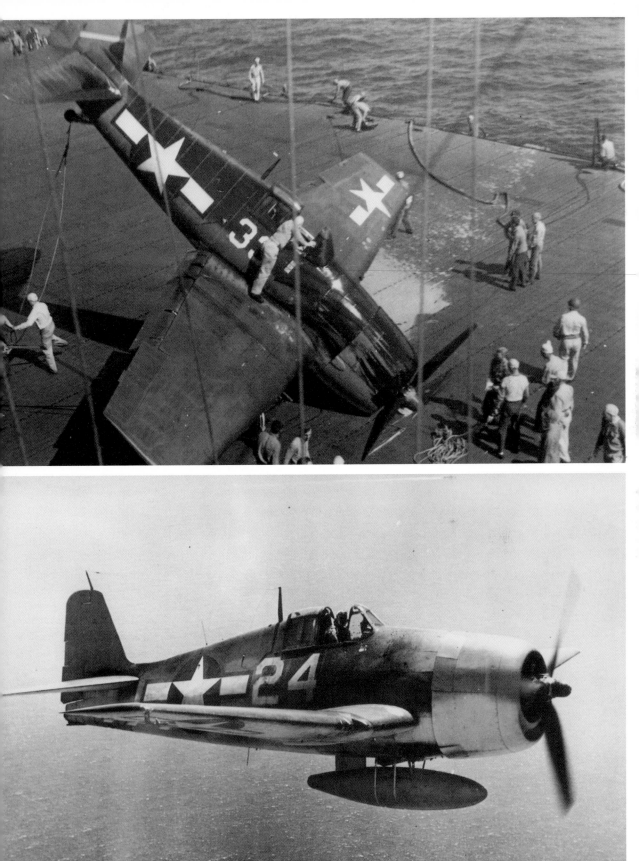

(*Above*) The Grumman F6F-3 Hellcats seen here in flight were low-wing monoplanes, in contrast to their F4F Wildcat predecessors, which where mid-wing monoplanes. The pilot of the Hellcat sat high on the top of the aircraft's self-sealing internal fuel tank, providing him with outstanding visibility. The F6F-3 was powered by a 2,000 hp engine. (*NNAM*)

(*Opposite above*) The follow-on version of the Grumman F6F-3 Hellcat was designated the F6F-5; a restored example is shown here. This second model of the aircraft had more armour. Not seen on this restored example of the F6F-5 are the wartime underwing attachments (hard-points) for bombs, or rockets. This was a feature not fitted to the F6F-3. (*Paul Hannah*)

(*Opposite below*) Grumman F6F-5 Hellcats on the flight deck of a US Navy carrier, some of which are being armed with rockets. The F6F-5 was designed with a strengthened tail section and trim tabs on its ailerons. Due to an increase in weight, the F6F-5, which was powered by the same 2,000 hp engine as mounted in the F6F-3, was a bit slower than the latter. (*NNAM*)

A flight deck dispatcher (currently known as the shooter) is signalling the pilot of a Grumman F6F-5 that his aircraft is about to be launched off the deck of a US Navy aircraft carrier. Visible in the photograph is the two-wire disposable catapult bridle that was employed to launch the plane from the ship. (NNAM)

(Opposite page) Looking down from the island of a US Navy aircraft carrier, is a Grumman F6F-5 Hellcat on the ship's flight deck, with its wings folded for storage. The F6F-5, as well as the F6F-3 Hellcat, had a length of 33 feet 7 inches long and a wingspan of 42 feet 10 inches. (NNAM)

A gear-up landing, as is seen with this Grumman F6F-5, can be due to either pilot error or mechanical failure, which can be a result of combat damage. Grumman built 6,436 units of the F6F-5. There were also 200 units built of a modified version of the F6F-5, intended for the photo-reconnaissance role, designated the F6F-5P. (RWP)

(*Above*) The Japanese turned to launching more aerial attacks during the hours of darkness late in the Second World War. To counter this threat, the US Navy fielded radar-equipped night-fighter versions of the Hellcat in late 1944. The letter 'N' for night fighter was added at the end of their designation. Pictured is an F6F-5N with the radar nacelle seen on the aircraft's starboard wing and a 20 mm cannon in each inner wing gun bay. (*NNAM*)

(*Opposite above*) The Vought F4U-1 Corsair shown here was first delivered to the US Navy at the end of October 1942. Notice the early style canopy that was later replaced with an improved visibility model. Banned from carrier use early on due to some design issues, the F4U-1 was restricted to being operated from land bases, primarily by US Marine Corps squadrons. (*NNAM*)

(*Opposite below*) Pictured is an early production US Marine Corps Vought F4U-1 Corsair, having its propeller being spun by a ground crewman prior to the engine starting, to avoid compression lock. The aircraft has the original type canopy with the extensive exterior bracing. The first US Navy F4U-1 Corsair land-based squadron to see combat was VF-17, nicknamed the 'Jolly Rogers'. (*NNAM*)

Mechanics work on a US Marine Corps Vought F4U-1 Corsair. Later production F4U-1s featured a number of design improvements, such as the new bubble-type canopy, as seen on the aircraft pictured, and are sometimes referred to as the F4U-1A. Grumman built 2,754 units of the F4U-1 and F4U-1A, with 533 passed on to America's wartime allies. (*NNAM*)

Ready to be launched from a US Navy carrier by catapult is a US Marine Corps Vought F4U-1A Corsair. It returned to daytime carrier duty beginning in January 1945. There were two related reasons for this: first, the danger posed to the fleet by the large number of Japanese kamikazes; second, a temporary shortage of F6F-6 US Navy Hellcat squadrons that could be assigned to the ships to counter the new threat. This meant that a number of US Marine Corps F4U Corsair land-based squadrons filled the void. (*NNMR*)

Pictured is a US Marine Corps F4U-1D, having just been launched from a carrier flight deck. Notice the disposable catapult launching bridle falling away. The Vought F4U-1D Corsair was configured as a fighter-bomber with under-wing attachments for a variety of ordnance, including bombs and rockets. A total of 1,375 units of the F4U-1D were built by Vought. (*NNAM*)

The last Vought-built version of their F4U Corsair series to see service during the Second World War was the F4U-4 shown here, however most were built after the conflict, with a total of 2,357 units completed. With a 2,100 hp engine, the aircraft could attain a maximum speed of 446 mph. (*RWP*)

(*Above*) Preparing for takeoff from a US Navy carrier, without benefit of a catapult, is a Vought F4U-1D Corsair fitted with an external drop tank. Twenty-four US Navy aviators and seventy US Marine Corps aviators became aces flying the F4U in the Pacific against the Japanese. The F4U series is credited with 2,140 enemy aircraft during the Second World War. (*NNAM*)

(*Opposite above*) Flying over the San Joaquin Valley of central California sometime in the 1950s is a Vought F4U-4 Corsair. The inverted gull-wing design of the aircraft was necessary due to the large 14-foot diameter propeller fitted to the plane. The inverted gull-wing design resolved this problem by placing the landing gear legs at the lowest point of the wings. (*NNRM*)

(*Opposite below*) Pictured is a restored Vought F4U-4 Corsair, fitted with a variety of dummy ordnance to replicate the appearance of the aircraft during the Korean War. It was employed as a fighter-bomber during that conflict as it was incapable of dealing with the enemy's faster jet-powered fighters, such as the Soviet-supplied MiG-15. (*Paul Hannah*)

To increase the wartime output of F4U Corsairs, the US Navy contracted with the Eastern Aircraft Division of General Motors to build a copy of F4U-1, known as the FG-1A, which lacked the folding wings of the Vought-built aircraft and was intended only for land-based squadrons. General Motors also built a copy of the Vought F4U-1D, labelled the FG-1D, with a restored example seen here. *(Paul Hannah)*

Shown on Peleliu in September 1944 is a US Marine Corps Goodyear-built FG-1D Corsair. In total, Goodyear built 2,302 units of the FG-1D model. Goodyear also built another model labelled the F2G-1, fitted with a 3,000 hp engine, but design problems and the end of the war resulted in only ten being built. *(NNAM)*

A restored Goodyear-built FG-1D Corsair is seen here with its wings folded. Besides Goodyear, the US Navy also contracted with the Brewster Aeronautical Corporation to build a copy of the F4U-1, labelled the F3A1; poor quality control led the US Navy to ban them from combat service, but 430 units out of 700 built still went to America's wartime allies. *(Christopher Vallier)*

A postwar-built version of the Vought F4U Corsair series employed by the American military was the F4U-5, with a restored example seen here. It was powered by a 2,850 hp engine. It was armed with four wing-mounted 20 mm automatic cannons, as had been the F4U-4C. *(Paul Hannah)*

(*Above*) The Vought F4U Corsair series was also adapted for the night-fighting role, with a wing-mounted radar pod. The final night-fighting version of the Corsair was based on the F4U-5 and referred to as the F4U-5N. An example is seen here during the Korean War, armed with air-to-surface rockets. (*NNAM*)

(*Opposite above*) Successful testing of an experimental version of the Vought F4U-5N, labelled the XF4U-6, resulted in the production of 111 units of a model for the US Marine Corps that was optimized for the ground attack role. It was designated the AU-1, and an example is seen here. It featured additional armour and could carry an underwing ordnance load of over 4,000 lbs. (*NNAM*)

(*Opposite below*) In June 1941, the US Navy gave Grumman the go-ahead to develop a twin-engine carrier fighter, the first ever ordered by the service. It would also have a secondary role as a fighter-bomber. Designated the F7F-1 Tigercat, it arrived too late to see combat during the Second World War. Pictured is a restored example of the two-man F7F-3 model of the aircraft. (*Christopher Vallier*)

By the time the Grumman F7F Tigercat series entered service in 1944, the US Navy had lost all interest in the aircraft as a carrier plane. Therefore, they were all passed on to the US Marine Corps as a land-based fighter-bomber. A small number of a two-man night fighter version of the aircraft, an example of which is seen here, did test fly off carriers, but were never assigned to the fleet in squadron service. (NNAM)

Another carrier fighter that reached the frontlines too late to see combat during the Second World War with the US Navy or US Marine Corps was the Grumman F8F Bearcat. Pictured is a restored example of the last model of the aircraft built, designated the F8F-2. It was intended as the replacement for the F6F-6 Hellcat. (Christopher Vallier)

On museum display is a Boeing FB-5, the first US Navy carrier fighter, although the aircraft pictured is in US Marine Corps markings. The initial production run of twenty-seven units was delivered directly from the factory to the US Navy's first aircraft carrier, the USS *Langley* (CV-1,) and made the first test flight from the ship. (*Christopher Vallier*)

Seen at an air show is a preserved Grumman F3F-2 fighter in US Navy markings. The F3F series also included the F3F-1 and the F3F-3. They were the last biplane fighters delivered the US Navy and armed with two machine guns. One hundred and forty-seven of all types were built. (*Paul Hannah*)

Taking part in an air show is a preserved FM-2 General Motors-built model of the Grumman-designed F4F Wildcat series of fighters. The first units of the FM-2 Wildcat were delivered to the US Navy in August 1943 and the last in September 1945, with a production run of 4,777 units. (*Christopher Vallier*)

Taking to the air during an air show is a preserved Grumman F6F Hellcat with its wheels being retracted into the wings. It was the successor to the Grumman-designed F4F Wildcat and the first American carrier-based fighter with operational parameters that was superior to the Mitsubishi A6M Zero, which had dominated the early part of the aerial war in the Pacific. (*Paul Hannah*)

Being demonstrated during an air show is a preserved Grumman-designed, General Motors-built TBM-3 Avenger torpedo-bomber in US Navy markings. When built by Grumman, its letter identifying code was 'TBF', and when built by General Motors it was labelled 'TBM'. In the rear upper fuselage is a power-operated ball turret armed with a .50 caliber machine gun to protect the aircraft from enemy fighters attacking from behind. *(Paul Hannah)*

Preparing for takeoff from a US Navy carrier during the Korean War is a Vought F4F-4 Corsair. Outclassed as an interceptor by post-war jet fighters, it served a useful role as a ground attack aircraft during the Korean War. A single US Navy pilot flying the Corsair became an ace during the conflict by shooting down five of the enemy's prop-driven aircraft. *(NNAM)*

On display at an airshow is a preserved North American FJ-4B Fury fighter in US Navy markings. The aircraft was based on the design of the North American F-86H Sabre fighter employed by the US Air Force, with some carrier-specific features such as arresting and catapult hooks. *(Christopher Vallier)*

The US Navy McDonnell F3H-2N Demon fighter shown here was developed specifically as a launch platform for the then brand-new Sparrow I air-to-air missile. The Demon itself was considered an underpowered second-rate aircraft by the pilots who flew it and was not missed when replaced. *(NNAM)*

Originally intended as a nuclear-armed carrier-launched bomber, the North American A3J/A-5 Vigilante spent the bulk of its long service career on US Navy carriers as a reconnaissance aircraft. As a specialized reconnaissance aircraft it was designated the RA-5C. (*DOD*)

This picture of a McDonnell Douglas A-4E Skyhawk was taken on 21 November 1967 and shows the aircraft en route to a target in North Vietnam. The plane is from US Navy Attack Squadron 164 (VA-164) and is being flown by the squadron's executive officer, Commander William F. Span. (*DOD*)

In this 1967 picture taken on a US Navy carrier flight deck is a Vought RF-8 photo reconnaissance Crusader ready to be launched, with another waiting its turn behind it. The one-man aircraft was powered by a single engine. It proved an extremely difficult aircraft to fly and had a very high accident rate during its service career. (NNAM)

(Opposite page) Shown in flight are two Vought A-7 Corsair IIs that resulted from a 1962 US Navy request for a replacement for the A-4 Skyhawk. Unlike the Vought F-8 Crusader on which it was derived, the Vought A-7 Corsair II was designed from the beginning as a ground attack, or strike aircraft. (DOD)

A picture taken on 8 November 1961 of a McDonnell Douglas F4H-1 Phantom II of US Navy squadron VF-74 assigned to the USS Forrestal (CVA-59). A total of 1,264 units of this twin-engine, two-seat aircraft were built in a wide variety of models for the US Navy and US Marine Corps. (NNAM)

Seen in a steep climb is a Grumman F-14 Tomcat, which came in three models. Clearly evident are the aircraft's variable-geometry wings designed for both speed and greater stability in flight. The Tomcat first saw action on 19 August 1981 over Libya when two F-14s shot down two Soviet-supplied Libyan Air Force Su-22 ground attack aircraft that tried to engage them in combat. (DOD)

In the colours and markings of the famous US Navy Blue Angels Flight Demonstration Team are two early-model F/A-18 Hornets, now unofficially referred to as 'Legacy Hornets', as they are being replaced by a larger, and more capable version of the aircraft referred to unofficially as the 'Super Hornet'. The Blue Angels were formed in 1946 and have flown at different times almost all the US Navy fighters. (DOD)

With the conclusion of the Second World War, the number of Grumman F8F Bearcats ordered was dramatically curtailed. The final production version of the aircraft was the F8F-2 model, a restored example of which is seen here. The aircraft was pulled from US Navy service prior to the Korean War. *(Christopher Vallier)*

The US Navy understood early in the Second World War that the long-term future of aircraft technology lay with jet-powered planes. The service's answer to the low-powered jet engines of the time was a hybrid aircraft that was both prop-driven and jet-powered. This resulted in the ordering in December 1943 of 100 Ryan FR-1 Fireballs, an example is seen here trying to land on a carrier flight deck. *(NNAM)*

Pictured is a restored Ryan FR-1 Fireball. The first units of the aircraft arrived in US Navy service in March 1945, but showed up too late to see combat before the conflict ended. With the conclusion of the war, the contract with Ryan for the FR-1 was cancelled after only sixty-six units were built. The aircraft's small jet engine was located in the rear of the plane's fuselage. (*Paul Hannah*)

The first all jet-powered aircraft to be flown by the American military was the US Army Air Force's experimental Bell XP-59A Aircomet in October 1942. It was followed by the YP-59A in June 1943. The US Navy acquired two of the YP-59As for test purposes and designated them the YF2L-1, but no orders followed. Pictured is a production model of the YP-59A, labelled the P-59B, ordered by the US Army Air Force in 1943. (*US Air Force Museum*)

In the same time frame as the jet-powered Bell YP-59A Aircomet, Lockheed was pursuing their own jet-powered aircraft design. That design proved more successful and eventually evolved into the P-80 Shooting Star. The US Navy acquired three units in June 1945, designated the P-80A, one of which is seen here in August 1945. (NNAM)

The US Navy tested the suitability of the one-man Lockheed P-80A Shooting Star in 1946 for carrier operations, but found it wanting. However, it was recognized that it made an excellent land-based training aircraft. Eventually, a two-man model, designated the TO-2, seen here, was acquired by the US Navy. (NNAM)

(*Above*) On a US Navy carrier flight deck is a Curtiss SB2C-3 Helldiver folding its wings for storage. This particular Helldiver model appeared in service in 1944. Notice the perforated trailing edge flaps that were used as dive brakes by pilots of the SB2C-3 and subsequent models of the aircraft, to recover from dive-bombing attacks. (*NNAM*)

(*Opposite above*) As carrier operations were such an important part of the training of every US Navy pilot, Lockheed eventually came up with an improved two-man version of the TO-2 suitable for carrier training. It was designated the T2V-1 Sea Star, an example of which is seen here on a carrier flight deck. (*NNAM*)

(*Opposite below*) A picture dated January 1943 shows an overhead view of a US Navy Curtiss SB2C-1 Helldiver scout-bomber. The prototype of the aircraft first flew in December 1940. Typically, the US Navy would then extensively test the prototype to uncover any design shortcomings that would have to be corrected prior to acceptance into service. However, with war clouds looming over America, the SB2C-1 was ordered into full-scale production by the US Navy and US Army a month before the prototype flew. (*NNAM*)

CURTISS SB2C-4
3/4 RIGHT FRONT VIEW
OF FUSELAGE IN JIG.
CB-12210 LY 10-5-44

(*Above*) A Curtiss SB2C-4E Helldiver that has just landed on the flight deck of a US Navy carrier is nosing over with its propeller digging into the wooden planks that cover the ship's non-armoured flight deck. Beginning with the SB2C-1c model, the plane was armed with two wing-mounted 20 mm automatic cannons, one in either wing. (*NNAM*)

(*Opposite above*) Shown on a Curtiss wartime assembly line is the uncompleted metal fuselage of an SB2C-4 Helldiver. The flight crew was protected by armour and the plane had self-sealing fuel tanks. Stored in the fuselage area between the pilot's cockpit and the rear machine gunner/radioman's position was a small inflatable life-raft. (*NNAM*)

(*Opposite below*) Returning to their carrier from a mission over Japan in the final month of the war is a formation of US Navy SB2C-4 Helldivers. The two-man aircraft was powered by a 1,900 hp engine that provided it with a top speed of 295 mph. It had a maximum take-off weight of 16,616 lbs that could include 2,000 lbs of ordnance. (*NNAM*)

Shown here at an airshow is a restored Curtiss SB2C-5 Helldiver in wartime US Navy markings. Like all the aircraft in the SB2C series, it was of all metal construction, except for its fabric-covered elevators and rudder. The plane was 36 feet 8 inches in length and had a wingspan of 49 feet 9 inches. Aircraft height was 13 feet 2 inches. *(Loren Hannah)*

On museum display is a Brewster SB2A Buccaneer scout-bomber in wartime US Navy markings. It was powered by a 1,700 hp engine that provided it with a maximum speed of 274 mph. The two-man aircraft was 39 feet 2 inches long and had a wingspan of 47 feet. It was 15 feet 5 inches tall. *(Paul Hannah)*

The cockpit of a Curtiss SB2C-5 Helldiver is shown. To ease the workload on the aircraft's pilot, it was equipped with an autopilot. Despite this helpful feature, pilots who had flown the older Douglas SBD dive bomber generally preferred it over the newer Curtiss SB2C series because it was smaller, weighed less, and had a longer maximum range. (NNAM)

(*Above*) A US Navy Grumman TBF-1C Avenger torpedo-bomber is shown launching a Mk. 13 torpedo that weighed 2,216 lbs and was armed with a 600 lb warhead. The torpedo had a range of 6,300 yards at 33.5 knots and was carried within the aircraft's cavernous internal bomb bay. In lieu of the single large torpedo, the TBF-1C could carry an assortment of different sized bombs. (*NNAM*)

(*Opposite above*) Pictured is a Grumman TBF-1C Avenger torpedo-bomber with its long transparent canopy. The aircraft had a three-man crew, with the pilot in the forward cockpit, the radioman/bombardier in the middle of the fuselage, and the rear gunner, who was tasked with operating the electrically-powered ball turret, armed with a .50 caliber machine gun, located behind the canopy. (*NNAM*)

(*Opposite below*) A Grumman TBF-1C Avenger torpedo-bomber is pictured on the flight deck of a US Navy carrier in this 1 January 1944 dated photograph. It is fitted with underwing rocket-launching rails. Rockets were first fired in combat by US Navy aircraft in January 1944 and became standard for all carrier aircraft by October 1944. (*NNAM*)

Pictured with its wings folded is a US Navy Grumman TBF-1C Avenger torpedo-bomber. As the Japanese Navy's large warships disappeared from the Pacific late in the war, the US Navy's need for a torpedo-bomber lessened and the Avenger became primarily a horizontal bomber. Avengers in the Pacific accounted for 32 per cent of the bomb expenditure dropped by US Navy and US Marine Corps aircraft. (*NNAM*)

As the US Navy wanted Grumman to concentrate on the building of fighters during the Second World War, construction of the original TBF-1 Avenger torpedo-bomber and the follow-on TBF-1C was farmed out to the Eastern Aircraft Division of General Motors. Built by General Motors, they were respectively labelled the TBM-1 and the TBM-1C. Pictured is a TBM-1C that has come to grief on a US Navy carrier flight deck in April 1945. (*NNAM*)

The final version of the Grumman-designed Avenger torpedo-bomber was a General Motors-built variant designated the TBM-3. It came in a number of sub-variants with additional letters added to the end of their aircraft designation to identify them. Pictured at an airshow is a restored TBM-3E in US Navy wartime markings. (*Paul Hannah*)

Being waved off from a carrier flight deck during the Second World War is a General Motors built TBM-3 armed with underwing rockets. Air-to-surface rockets were originally envisioned by the US Navy as an anti-submarine weapon. However, their obvious suitability for other roles resulted in the US Navy adopting them for a wide range of missions. (*NNAM*)

(*Above*) The wartime-built Martin PB2Y series of Coronado seaplane patrol bombers had stabilizing floats which retracted in flight to become wingtips. Here we see what happens when one of those stabilizing floats fails for whatever reason. The crew of a Coast Guard launch has attached a rope to the aircraft's wing and is no doubt attempting to right it. (*NNAM*)

(*Opposite above*) Even before ordering the Consolidated PBY Catalina into production, the US Navy was exploring the possibility of even larger four-engine patrol-bomber seaplanes in 1935. The end result was the Martin PB2Y-3 Coronado. A total of 210 units of the aircraft were built by Martin, most were later upgraded with new engines and re-labelled as the PB2Y-5. The sole surviving example of the production run is seen here on display. (*NNAM*)

(*Opposite below*) The largest seaplane to see service with the US Navy in the Second World War was another Martin product, labelled the JRM-1 Mars, an example of which is seen here. Twenty of the aircraft were ordered by the US Navy in January 1945, however only five were completed when the Japanese surrendered in September 1945, resulting in the remaining seaplanes being cancelled. (*RWP*)

In the immediate postwar years, the four remaining JRM-1 Mars seaplanes, one having been destroyed in a crash in August 1945, were upgraded and re-designated the JRM-2 Mars, based on a single new-built unit completed in 1947. Pictured on the ground is an example of the JRM-2 Mars, with the marking of the Naval Air Transport Service, which existed from 1941 until 1948. *(RWP)*

The US Navy decided in 1938 that they needed a new observation floatplane to replace the Curtiss SOC Seagull onboard cruisers. Once again, the US Navy chose a Curtiss product they labelled the SO3C-1 Seamew. The prototype is shown here. The first production unit was delivered in 1942. Due to a number of design issues, the SO3C-1 and subsequent models were quickly pulled from service. *(National Archives)*

After the failure of the Seamew, in 1942 the US Navy decided that they needed a new observation floatplane to replace it, and the pre-war designed Vought OS2U Kingfisher Curtiss therefore began work on an aircraft that eventually came off the production line in October 1944, designated the SC-1 Seahawk, an example of which is shown here. An improved model was labelled the SC-2. *(NNAM)*

Descended from a pre-war passenger plane is the Lockheed PV-2 Harpoon maritime patrol plane. It is shown here in US Navy wartime markings as a museum display. The earlier version of the aircraft was known as the PV-1, but had not been assigned the name Harpoon. The PV-2 had a six-man crew, was 51 feet 5 inches long and had a wingspan of 65 feet 6 inches. *(NNAM)*

A restored Lockheed PV-2 Harpoon maritime patrol bomber is shown at an airshow in the markings of a wartime US Navy aircraft. The aircraft was capable of carrying an ordnance load of 4,000 lbs in an internal bomb bay, and eight underwing 5-inch rockets, out to a range of 950 miles. The PV-2D model was armed with eight forward-firing machine guns. (*Loren Hannah*)

A long line of B-24D Liberator heavy bombers is seen in this picture of a Consolidated assembly line. Besides their intended role as long-range strategic bombers, they were employed by the US Army Air Forces in 1942 as long-range maritime patrol bombers. This was a role that was later turned over to the US Navy in 1943, along with 977 units of various versions of the B-24, which the US Navy designated the PB4Y-1. (*US Air Force Museum*)

Pictured is a US Navy Consolidated PB4Y-1. In service over the Atlantic Ocean, PB4Y-1-equipped squadrons accounted for ninety-seven German submarines during the Second World War. The aircraft was equipped with an air-to-surface radar unit intended to detect enemy submarines on the surface, as well as enemy surface ships. (*NNAM*)

Pleased with the Consolidated PB4Y-1, the US Navy contracted with the firm to build a dedicated version of the B-24 Liberator as a maritime patrol bomber. It appeared in service in late 1944 and was designated the PB4Y-2 Privateer, an example of which is seen here. Consolidated built 739 units of the PB4Y-2 for the US Navy, most after the conclusion of the Second World War. (*RWP*)

The Consolidated PB4Y-2 Privateer, pictured here, could easily be distinguished from the B-24 Liberator series, including the PB4Y-1, by its single tall vertical stabilizer. This was in place of the twin-tail vertical stabilizer arrangement on all the various versions of the B-24 Liberator series. Note also the PB4Y-2's two dorsal turrets, and twin .50 caliber machine gun-equipped waist stations. The PB4Y-2 could carry an ordnance load of 12,800 lbs within its internal bomb bay. (*NNAM*)

Shown at an airshow is a restored North American PBJ-1 Mitchell medium bomber in US Marine Corps markings. The US Navy acquired 706 units of the PBJ-1 in various models for use by the US Marine Corps. The late production units of the PBJ-1 series were constructed without the upper fuselage machine gun turret. (*Christopher Vallier*)

The US Army Air Force made widespread use of the Martin B-26 Marauder medium bomber during the Second World War, seen here in flight. Early production units were employed in non-combat roles, such as target towing and utility purposes by both the US Army Air Force and the US Navy during the Second World War. (*US Air Force Museum*)

During the last year of the Second World War, the US Navy acquired two Douglas A-26C attack/light bombers from the US Army Air Force. This was followed by 150 additional units in the immediate postwar period. In US Navy service it was labelled the JD-1, with an example shown here. It was used as a target tug and utility aircraft. (*NNAM*)

Five days after the Japanese attack on Pearl Harbor, the US Navy formed the Naval Air Transport Service, the main-stay of which became the Douglas C-47 Skytrain. In US Navy and US Marine Corps service it was designated the R4D. Pictured at a museum is an R4D series aircraft that was upgraded in the early postwar period by Douglas for the US Navy and re-designated as the C-117D. *(NNAM)*

During the Second World War the US Navy acquired for the US Marine Corps from US Army Air Force contracts 160 units of the Curtiss C-46 Commando. The US Navy labelled the twin-engine transport the R5C-1. Pictured is a restored example of a C-46 Commando, in US Army Air Force markings. *(Christopher Vallier)*

Taken into service by the US Army Air Corps prior to America's official entry into the Second World War was the Douglas C-54 four-engine Skymaster transport. The US Navy acquired 183 units between 1943 and 1945 and designated it the R5D series, an example of which is seen here. (*NNAM*)

In anticipation of training a large number of pilots, the US Army Air Corps awarded a contract to Cessna for a version of its Model T-50 twin-engine passenger plane. The US Navy saw its usefulness and picked up sixty-seven units of the aircraft, which they designated the JRC-1, an example of which can be seen here on display. (*NNAM*)

The US Navy assigned the job of helicopter development to the US Coast Guard during the Second World War. What eventually showed up in service late in the conflict were twenty-five units of a helicopter, designated the HNS-1, an example of which is seen here with a line of F4U Corsairs in the background. (*NNAM*)

Chapter Three

Cold War Aircraft

With the beginnings of the Cold War in 1948, nobody could conceive of any future wars between the two new super powers, the United States and the Soviet Union, which did not involve the use of nuclear weapons. This left the US Navy as a service that had no way of participating in the next global conflict, as it then lacked any long-range carrier aircraft capable of delivering the 10,000 lbs nuclear bombs of the day. The largest ordnance payload for a Second World War single-engine US Navy carrier-based aircraft was only 2,000 lbs.

As a stop-gap measure to prove that a long-range, multi-engine, nuclear-capable aircraft could be launched from a carrier, the US Navy pulled twelve of its brand-new Lockheed P2V-2 land-based Neptune maritime patrol planes from service and modified them into one-way, long-range, nuclear-capable, carrier takeoff bombers that could attack the Soviet Union. They were powered by two prop-driven engines, supplemented by two jet pods for help when being launched from land bases. In their new role they became the P2V-3C.

Dedicated Nuclear-Capable Bombers

Even as the first squadron of nuclear-capable Neptune bombers was being formed, the US Navy had a replacement aircraft ready to come off the production line. That plane was the North American AJ Savage. The letter 'A' in the aircraft prefix designation code stood for 'attack', a term introduced in 1946, replacing the pre-Second World War and wartime use of the letter 'T' for torpedo, and the letter 'B' for bomber. The letter 'J' in the aircraft prefix code was the builder's letter code.

The various versions of the nuclear-capable Savage, the AJ-1 and AJ-2, were in service with the US Navy from 1950 until 1958, with 140 units built. There was also a photo-reconnaissance version of the aircraft designated the AJ-2P, which the AJ-2 model was based on. Like the nuclear-capable Neptune, the nuclear-capable Savage had two large prop-driven engines, but supplemented by a small jet engine to aid in launching it from a carrier flight deck and providing it with dash speed when being chased by enemy aircraft.

The Savage's successor was the nuclear-capable, jet-engine-powered Douglas A-3A Skywarrior bomber that entered US Navy service in 1956. It was at the time the largest and heaviest aircraft to operate off a carrier. The initial production model, the A3D-1 was soon superseded by an improved version, designated the A3D-2 that first reached the fleet in 1957. A total of 283 units of the Skywarrior were built. The aircraft was relabelled the A-3 Skywarrior in 1962 as part of the Tri-Service Aircraft Designation System mandate.

For the sake of brevity, it will be assumed in the remainder of the text that the reader will understand that all US Navy and US Marine Corps aircraft were relabelled in 1962 with new designations to conform to the Tri-Service Aircraft Designation System mandate imposed by the US Congress.

The Skywarrior was replaced in 1961 by another jet-powered, nuclear-capable bomber, the North American A3J Vigilante. A total of 156 units were built. Less than a year after the Vigilante entered service it was relabelled the A-5 Vigilante. At the same time, its role of delivering nuclear ordnance to the Soviet Union was transferred to the US Navy's new inventory of nuclear-powered ballistic missile submarines (SSBNs) equipped with the long-range, nuclear-armed, Polaris missile system.

Nuclear bombs did not disappear from the arsenal of possible weapons carried by carrier aircraft in 1962. Rather they had progressively gotten small enough to be carried by existing attack aircraft. This did away with the need for large specialized bombers such as the Savage, Skywarrior, and Vigilante.

New Roles

Not wanting to waste perfectly fine aircraft upon passing the long-range nuclear strike role against the Soviet Union to its submarines, the US Navy converted all of its existing, and yet to be delivered A-5 Vigilante nuclear-capable bombers into photo-reconnaissance aircraft and assigned them the designation RA-5C. The letter 'R' in the aircraft's prefix designation code represented reconnaissance. In US Navy service in this latter role the Vigilante lasted until 1979.

The Skywarrior was converted into a photo-reconnaissance aircraft, designated the RA-3B or an inflight tanker designated the KA-3B. The letter 'K' in the aircraft designation code represented tanker. During the early part of the Vietnam War the Skywarrior was pushed into new roles, such as performing conventional bombing missions, as well as being configured as an electronic reconnaissance plane (EA-3B), with the letter 'E' standing for special electronics installation. The aircraft was phased out of US Navy service in 1991.

First-Generation Fighters

The first of the postwar carrier-capable jet fighters was the McDonnell FH-1 Phantom. A total of sixty-two were built and they served from 1947 until 1954 with both the US Navy and US Marine Corps. At the same time the Phantom entered service, the North American FJ-1 Fury showed up. Only thirty-one were built and it served with the US Navy until 1953.

On the heels of the Phantom and the Fury came other first-generation jet fighters, such as the McDonnell F2H Banshee, a larger improved fighter based on the design of their earlier Phantom. A total of 895 were built and it was in service with the US Navy and US Marine Corps from 1948 until 1961. In the same time frame, the Vought F6U Pirate began undergoing testing in 1946, however it failed to meet expectations and was cancelled in 1950, after only thirty-three were built.

Other first-generation carrier jet fighters included the Grumman F9F Panther, the Douglas F3D Skynight (intended as an all-weather dedicated night fighter) and the Vought F7U-1 Cutlass. A total of 1,382 units of the Panther were built and it served from 1949 until 1958 with the US Navy and Marine Corps. It would be numerically the most important US Navy jet during the Korean War. Two hundred and sixty-five units of the Skynight were built and it served from 1950 until 1970 with the US Navy and US Marine Corps. In 1962, the Skynight was re-labelled the F-10. Three hundred and twenty units of the Vought F7U-1 Cutlass were constructed. The plane itself was considered difficult to fly and underpowered, which resulted in a great many crashes. It would last in service only from 1951 until 1959.

The Banshee, Panther, and Skynight saw combat during the Korean War, the first two as ground attack aircraft, and the latter in its intended role as a night fighter. The Skynight also saw action during the Vietnam War as an electronic warfare bird in its EF-10 guise, the only US Navy fighter to see service in both conflicts. The other first-generation fighters were not employed during the Korean War or Vietnam War for a number of different reasons, including design faults and operational performance limitations.

Second-Generation Fighters

Key design features of the second-generation carrier jet fighters were their super-sonic speed, new radar, air-to-air missiles, and swept wings. The swept wing design was based on German jet aircraft research during the Second World War. All the first-generation carrier jet fighters were sub-sonic and had straight wings, except the Cutlass that had a unique swept-wing design.

Second-generation carrier fighter jets included various models of the McDonnell F3H Demon, the Douglass F4D Skyray, and the Grumman F9F Cougar, which was a swept-wing version of the first-generation Grumman F9F Panther, with a more powerful engine.

The Demon, of which 559 units were built, saw service between 1956 and 1964. The Skyray, of which 422 units were built, entered service in 1956 and it lasted until 1964. A total of 1,392 units of the Cougar were constructed and it served from 1952 until 1974 with the US Navy and US Marine Corps mainly as a ground-attack aircraft.

In 1962, the Demon was re-labelled the F-3, the Skyray the F-6, and the F9F-6 the F-9. Of the three aircraft, only four training versions of the latter, designated the TF9J Cougar, saw action during the Vietnam War, as command and control aircraft for the US Marine Corps. The letter 'T' in the aircraft's designation stood for trainer.

Another second-generation US Navy jet was the North American FJ-2/3 Fury series, a navalized version of the North American F-86 Sabre fighter series, employed by the US Air Force with great success during the Korean War. The US Navy and US Marine Corps took 741 units of the Fury into service, starting in 1954. It lasted in US Navy service until 1956 and with the US Marine Corps until 1962.

A second-generation design failure was the Grumman F-11F Tiger that entered the US Navy inventory in 1954; 199 were constructed. However, it lasted only four years on carriers, due to unsurmounted design problems, before being relegated to training duties between 1961 and 1967. It served with the US Navy's demonstration team, the Blue Angels, until 1969. In 1962, the F-11F Tiger became the F-11 Tiger.

In the late 1950s, the last of the second-generation jet fighters appeared in US Navy and also US Marine Corps service. The first was the Vought F-8U Crusader in 1957, which became the F-8 Crusader in 1962. The aircraft saw combat during the Vietnam War as both a fighter and as a light attack aircraft. It lasted in US Navy service until 1987 in the photo-reconnaissance variant, the fighter models being retired from US Navy and US Marine Corps service by 1976. A total of 1,219 units of the Crusader were built.

Third-Generation Fighters

The replacement for the F-8 Crusader in US Navy and US Marine Corps service was the McDonnell F-4 Phantom II series that first appeared on carriers in 1960. In 1962, the plane was designated the F-4B Phantom II. The US Air Force was so impressed by the F-4B Phantom II that they adopted a land-based version for their own use in 1964 and designated it the F-4C Phantom II.

In 1967, the McDonnell Aircraft Corporation acquired Douglas Aircraft, and the combined firms became known as the McDonnell Douglas Corporation. As their business faded in the 1990s with the end of the Cold War in 1991 and a dramatic cut in defense funding, the firm was acquired by the Boeing Corporation in 1997, and the corporate name McDonnell Douglas disappeared.

Like all the first- and second-generation jet fighters that preceded it into service, the Phantom II was originally intended solely for the role of interceptor, but was later pressed into other roles, such as ground attack during the Vietnam War, and photo reconnaissance and suppression of enemy air defenses. There were countless models of the Phantom II placed into service during its long service career. By the time the production run of the Phantom II ended in 1981, a total of 1,264 units had been acquired by the US Navy and the US Marine Corps, with hundreds more exported to US allies.

A Design Dead End

Prior to the entry of the Phantom II into service, the US Navy began exploring concept development of a more capable Fleet Air Defense fighter to protect carrier battle groups from long-range Soviet bombers. The USAF, at the same time, desired a new deep strike aircraft. Secretary of Defense McNamara forced both services to study purchasing a common aircraft as a way to save money, calling the program the Tactical Fighter Experimental (TFX). In the end, the services failed to reconcile their requirements.

The US Navy version of the TFX program was the General Dynamics F-111B. General Dynamics (GD) had entered into the military aircraft field in 1953 by acquiring ownership of Convair, which then became the Convair Division of GD. In 1994, GD sold its Convair Division to Lockheed. The corporate name Convair disappeared in 1996.

The first test flight of the F-111B took place in 1965. Additional testing of the aircraft was a disappointment to many in the US Navy, who felt it was too heavy and underpowered for operating off carriers. The project was cancelled in 1968 with just eight of the aircraft being built, however key elements of the TFX were used in the next generation US Navy fighter. The US Air Force did take a different version into service as the F-111 Aardvark in 1967 and employed it as a deep strike, electronic warfare, and tactical nuclear delivery aircraft until 1998.

Fourth-Generation Fighters

The fourth generation of supersonic jet fighters acquired by the US Navy began with the Grumman F-14 Tomcat. The twin-engine, two-seat aircraft entered US Navy service in 1974 and was the replacement for the McDonnell Douglas F-4 Phantom II. It came in two major versions during the Cold War; the 'A' and 'B' variants. The US Navy took into service 478 F-14A Tomcats beginning in 1970. The improved F-14B Tomcat appeared in 1987 and consisted of thirty-eight new-built aircraft and forty-eight A-model Tomcats brought up to the B-model standard.

The Tomcat was not adopted by the US Marine Corps because it had not been configured in the original model for a secondary ground attack role. Instead, the US Marine Corps acquired the McDonnell Douglas/British Aerospace AV-8A Harrier. A handful of F-14As still serve in the Iranian Air Force.

The second of the fourth-generation fighters was the McDonnell Douglas F/A-18 Hornet. The prefix 'F/A' reflects the Hornet's dual role as both a fighter and a strike aircraft. The term 'strike' is employed by the US Navy to describe a multi-mission aircraft and was adopted by the US Navy for the Hornet, in lieu of the older term 'attack', which had been adopted in 1946 and was meant for specially-designed aircraft intended primarily for the ground attack role.

In 1983, the US Navy renamed its existing Hornet-equipped units as 'strike fighter squadrons', the previous name being 'fighter attack squadrons'. A somewhat older, now generic term, for fighters capable of a ground attack role is 'fighter-bomber'.

The F/A-18 Hornet appeared in US Navy and US Marine Corps squadron service in 1983. It was the replacement for a number of aircraft including the A-6 Intruder, the A-7 Corsair, and the F-4 Phantom. The Hornet first flew combat missions during the 1986 attack on Libya. It also saw action during America's two wars with Iraq, known as Operation Desert Storm in 1991, and Operation Iraqi Freedom in 2003. It also flew combat missions over Afghanistan.

The F/A-18 originally came in a single-seat model, designated the F/A-18A and a two-seat model designated the F/A-18B. Approximately 400 units of the F/A-18A and F/A-18B entered into US Navy and US Marine Corps service. Beginning In 1987, the A and B models of the Hornet were progressively replaced by the much more capable C and D versions. Today, besides being employed as a training aircraft, the two-seat versions of the Hornet retains the ability to perform combat roles; in particular, the US Marine Corps uses the two-seat F/A-18D for the demanding night attack mission. Production of the F/A-18C model ended in 1999 and the D-model in 2000.

Due to a shortage of the F/A-18C Hornet version, a least one US Navy squadron flew the F/A-18A model into the early part of the twenty-first century. To keep the 'A' model of the Hornet a viable combat aircraft, sixty-one units were upgraded. Positive results achieved with the upgrading process resulted in fifty-four units being subsequently upgraded to an even more capable standard, which brought them up to the same operational capabilities as the F/A-18C model.

As of 2014, the US Navy and Marine Corps inventory of the A, B, C and D models of the Hornet consists of approximately 600 units. Current plans call for twenty-five units of the F/A-18C to be upgraded to a version referred to as the F/A-18C plus that will replace twenty-five aging units of the upgraded F/A-18A. Due to delays in the operational deployment of the post-Cold War replacement for the A, B, C and D model Hornets, it is envisioned that 150 upgraded units of the aircraft will remain in service for a number of years.

Attack Planes

Before the Hornet appeared on US Navy carriers, there were a number of aircraft dedicated to the attack role that flew from carriers during the postwar years. The first of these was the Douglas Skyraider that came in numerous models, AD-1 through AD-7, with sub-variants of each model, not all being employed by the US Navy and US Marine Corps. In 1962, the last three models of the Skyraider built were re-labelled. The AD-5 became the A-1E, the AD-6 became the A-1H, and the last model, originally designated the AD-7, was re-labelled as the A-1J.

The Skyraider was a prop-driven aircraft originally designed during the Second World War, but did not begin appearing on US Navy carriers until 1949. It saw service in both the Korean and Vietnam Wars before being retired by the US Navy in 1968. Total production of the Skyraider numbered 3,180 units, with many being employed by the US Air Force during the Vietnam War, but not the Korean War.

The Skyraider was not the only prop-driven specialized ground attack aircraft adopted by the US Navy in the early postwar years. There was the Martin AM-1 Mauler, but it did not live up to expectations and was in service only from 1948 until 1953, before the US Navy withdrew it in favor of the better performing Skyraider. Only 151 units of the Mauler were built.

The jet-powered follow-on to the Skyraider in the light attack aircraft category was the subsonic Douglas Skyhawk series that first showed up in US Navy and US Marine Corps service in 1956. It eventually served in a variety of versions. The pre-1962 designation system labelled them the A4D-1, the A4D-2, A4D-2N, and the A4D-5. In 1962, they became respectively the A4-A, the A-4B, A-4C, and the A-4E.

Appearing in US Navy and US Marine Corps service after 1967 was the A-4F model of the Skyhawk that can be easily identified by the upper fuselage hump pod that contained additional avionics. One hundred units of the A-4C were later rebuilt to the A-4F model standard and designated the A-4L. They served only with US Navy Reserve squadrons. The US Marine Corps employed 158 units of the aircraft designated A-4M Skyhawk, that had a more powerful engine and improved avionics.

The last production unit of the A-4M was delivered to the US Marine Corps in 1979, with the Skyhawk series being withdrawn from US Marine Corps service in 1998, and US Navy use in 2003. A total of 2,960 units of the aircraft were built, with over 550 being two-seat trainers.

The eventual replacement for the Skyhawk on US Navy carriers in 1966 was the Vought A-7 Corsair II. It saw combat in the Vietnam War and remained in service long enough to be employed during Operation Desert Storm in 1991. It was retired soon after that Middle Eastern conflict. In total, the US Navy acquired 997 units of the Corsair II, with 60 being two-seat trainers, designated the TA-7C. The Corsair II was not adopted by the US Marine Corps, which preferred to stay with the Skyhawk, until it could be replaced by the F/A-18 Hornet. The Corsair II also served with the USAF and several US allies.

Supplementing the light attack Skyhawk and Corsair II, beginning in 1963, and eventually replacing them on US Navy carriers was the Grumman A-6 Intruder, classified as a medium attack aircraft. The Intruder was an all-weather aircraft that could also operate at night. Its baptism in combat was the Vietnam War, with the initial model labelled the A-6A. Later versions included the A-6B, A-6C, and the final model, the A-6E that entered service in 1970.

Over 700 units of the Intruder eventually entered service with the US Navy and US Marine Corps. The latter retired their Intruder inventory in 1993 and the US Navy in 1996. It was the last dedicated attack aircraft in US Navy and US Marine Corps service.

A variant of the Intruder still in service is the EA-6B Prowler, which is an electronic-warfare (EW) aircraft intended to degrade enemy air-defense systems by jamming their electronic signals or killing them with anti-radiation missiles. The aircraft first entered service in 1971 with the US Navy and US Marine Corps. It will be retired from US Navy service in 2015, but retained by the US Marine Corps until 2019.

There was also an aerial refuelling version of the A-6 Intruder, designated the KA-6D. It could carry over 3,200 gallons of jet fuel that was transferred to other air-craft by hose-and-drogue pods. In total, ninety units of the KA-6D were placed into service by converting older model Intruders to the new role. Due to age-related fatigue problems, the aircraft is no longer in service with the US Navy.

Seaplanes

With the cancellation in 1949 of the first planned super carrier, the USS *United States* (CVA-58), the US Navy feared that it would have no way to deliver nuclear weapons onto targets in the Soviet Union. It quickly looked at a number of options, one of them a long-range, nuclear-capable seaplane bomber. What eventually resulted from this concept was the four-engine, jet-powered Martin P6M-2 Seamaster that appeared in early 1959. However, serious design problems plagued the aircraft and it was cancelled in late 1959, after only eight were built.

In 1948, there was an unfounded concern that supersonic jet aircraft might not be able to operate off the US Navy's carriers. This led the US Navy to believe that a supersonic seaplane fighter might fill the void. Convair provided the US Navy with an aircraft they thought would meet their needs, labelled the XF2Y-1 Sea Dart.

The first prototype of the Sea Dart flew in January 1953, with unimpressive results. The following year, a test model of the Sea Dart disintegrated in flight, killing the pilot. At the same time it was shown that supersonic aircraft *could* fly off carriers and the Sea Dart program was quickly cancelled.

A number of pre-war and wartime seaplanes saw early Cold War service with the US Navy before being retired. The only Cold War-era seaplanes acquired by the US Navy were both twin-engine, prop-driven aircraft; the Grumman JR2F-1 Albatross in 1949, and the Martin P5M Marlin in 1951. Whereas the Albatross was primarily a search and rescue aircraft, the Marlin was a maritime patrol aircraft, with a secondary ASW (Anti-submarine warfare) role. The Marlin was derived from the pre-war Martin PBM Mariner employed by the US Navy during the Second World War.

In 1962, the Albatross was re-designated as the HU-16 and the Marlin became the P-5. The last flight of a US Navy Albatross occurred in 1976 when it was flown to a museum for display. In Coast Guard service, the Albatross lasted until 1983 before being replaced by more modern aircraft. The latter remained in US Navy service until 1967 and did see action during the early part of the Vietnam War. A total of 449 units of the Albatross were built, with 285 units of the Marlin constructed.

Land-Based Maritime Patrol Aircraft

In the immediate postwar years, the Soviet Navy began to greatly expand its inventory of submarines, initially diesel-electric and then nuclear-powered. Their job was the destruction of NATO resupply convoys, US Navy carriers and their supporting ships, in the case of a third World War. To counter this very serious threat, the US Navy looked at two different land-based maritime patrol aircraft capable of an ASW role during the Cold War.

The land-based maritime patrol aircraft that the US Navy initially chose in 1947 was the Lockheed P2V Neptune. Power for the aircraft was provided by two prop-driven engines. It was built in seven major models, labelled P2V-1 through P2V-7. New designations appeared in 1962, with the P2V-4 through P2V-7 becoming respectively the P-2D, P-2E, P-2F, and the P-2H. Beginning with the P-2E version, the aircraft was fitted with two jet engine pods to assist in takeoffs.

The Neptune remained in US Navy service until 1984 and came in a wide range of models, including a version configured as a nuclear-capable bomber as previously mentioned in the text. In total, 1,181 units of the Neptune were constructed. The Neptune was also used by the US Army in small numbers during the Vietnam War for the electronic surveillance role.

The US Navy replacement for the P-2 Neptune series was the Lockheed P-3 Orion series that first appeared in 1962, and has appeared in a large number of models over the years, with 734 units completed. It is a four-engine prop-driven aircraft still in service with the US Navy. The P-3 Orion was based on the design of the Lockheed L-188 passenger plane. There is also an ELINT (Electronic Intelligence) version of the aircraft in US Navy service, labelled the EP-3E ARIES (Airborne Reconnaissance Integrated Electronic System).

Carrier ASW Aircraft

Reflecting the serious threat posed to US Navy carriers by Soviet Navy submarines during the Cold War, a number of aircraft configured for the ASW role operated off US Navy carriers during that time period. These included wartime and early postwar prop-driven aircraft modified for the role, followed later by specially-designed aircraft configured for the job.

Following the wartime aircraft modified for the ASW role came the postwar, prop-driven, Grumman AF Guardian. Originally intended as a torpedo-bomber it was decided to configure it for the ASW role. It worked in pairs, with the plane equipped with the detection gear labelled the AF-2W, and the second Guardian armed with weapons labelled the AF-2S. These aircraft entered service in 1950 and remained in carrier use until 1955.

The two-aircraft ASW combination was far from the optimum arrangement for carriers, which were always hard-pressed for space. The solution arrived in the form of a new prop-driven, twin-engine aircraft especially designed for the ASW role, which combined the detection gear and weapons needed to destroy enemy submarines in a single airframe. That aircraft was the Grumman S2F Tracker, which appeared in service in 1954, as the replacement for the Guardian.

The original production version of the Tracker was designated the S2F-1 and the last model constructed the S2F-3. A total of 1,284 units of the aircraft were built. The plane was re-designated in 1962, with the S2F-1 becoming the S-2A and the S2F-3 becoming the S-2D. All the various models of the Tracker were withdrawn from US Navy service by 1976.

The replacement for the Tracker was the Lockheed S-3A Viking that began appearing on carriers in 1974. A total of 186 were built. Between 1987 and 1994 a total of 119 units of the S-3A were upgraded to prolong their useful service life and became the S-3B.

Unlike the Tracker that was prop-driven, the Viking was powered by two jet engines. With the end of the Cold War in 1991, the pressing need for ASW aircraft on US Navy carriers diminished. The Viking was re-configured for the surface warfare role and as a carrier-onboard-delivery (COD) plane. With the arrival of newer aircraft to fulfill those roles, the Viking was retired from service in 2009.

Carrier and Land-Based AEW Aircraft

The need for an airborne early warning (AEW) aircraft was brought home to the US Navy during the last few months of the Second World War in the Pacific when large numbers of kamikazes began overwhelming the fleet's existing air-defense system. To rectify this problem, a number of options were explored. One involved mounting a radar system in a suitable aircraft that once aloft, could provide over-the-horizon radar coverage. The aircraft chosen for that role was the Grumman-designed TBM Avenger torpedo-bomber, re-designated in its new role as the TBM-3W.

The TBM-3W showed up too late to see action in the Second World War, but did lay the groundwork for successor AEW aircraft. Following in the footsteps of the TBM-3W, the US Navy took into service a number of US Army Air Force B-17 four-engine, land-based bombers, configured as AEW aircraft, and designated them as the

PB-1W. Besides the radar, they carried aloft a fighter-director team to assist in vectoring US Navy fighters onto threats identified by the plane's radar system.

The PB-1W was soon replaced in the AEW role by the conversion in 1954 of the four-engine, prop-driven, land-based Lockheed Constellation passenger plane. In its new role, it became the PO-1W Warning Star and in 1962 the EC-121 Warning Star, with the last Constellation variant remaining in US Navy service until 1982. The US Navy also employed fifty units of a cargo/passenger version of the Super Constellation originally labelled the R7V-1. It became the C-121G in 1962 and lasted in service until the 1970s.

Still, the US Navy needed a newer generation dedicated carrier-launched AEW aircraft. For a time, they employed a version of the Skyraider in that role, designated the AD-4W. One hundred and sixty-eight were built, with the US Navy eventually transferring fifty to the Royal Navy. As more capable AEW aircraft entered the US Navy inventory, the AD-4W was assigned to ASW duties and remained in service until 1965.

The US Navy replacement for the AD-4W Skyraider was the twin-engine, prop-driven Grumman WF-2 Tracer, which entered service in 1958. It was a variant of the Grumman S-2 Tracker ASW aircraft. The WF-2 Tracer became the E-1B Tracer in 1962. A total of eighty-eight units of the Tracer were built with the last retired from service in 1976. From the Tracer was derived the Grumman C-1 Trader COD aircraft that was in service between 1958 and 1988. The Trader became the TF-1 in 1962.

The Tracer was replaced by the Grumman twin-engine, prop-driven, E-2 Hawkeye series that initially entered service in 1964. Beset by early design problems, the aircraft has been continuously upgraded over the decades to improve its effectiveness and remains today an important part of every US Navy carrier. One hundred and thirty-three units of the Hawkeye have been built, in three versions; labelled E-2A, E-2B, and E-2C. All of the 'A' models of the aircraft were retired in 1967. With the miniaturization of electronics, the E-2 evolved from its early role as AEW, to include battle management and command and control.

Helicopters

The first helicopter in immediate postwar use by the US Navy was the wartime-built Sikorsky HNS-1 and the very similar Sikorsky HOS-1, developed by the US Coast Guard. In total, Sikorsky built 131 units of the HNS-1 and HOS-1 for a variety of customers, not just the US Navy. Some served with the Royal Navy.

Positive feedback from the testing of the wartime-built Sikorsky helicopters, as well as postwar Piasecki helicopter designs such as the HRP-1 Rescuer, of which the US Navy acquired twenty, resulted in improved models being acquired by the

US Navy, designated the HUP-1 and HUP-2 Retrievers. Piasecki built a total of 339 units of the Retrievers for several customers, including the US Navy.

In US Navy service, the Retrievers served in both the utility and search and rescue (SAR) roles, as well as performing an ASW role. They remained in US Navy service until 1964. In 1962, the HUP-1 and HUP-2 respectively became the UH-25B and the UH-25C.

The first Sikorsky Cold War-era helicopter acquired by the US Navy and the US Marine Corps were eighty-eight units designated the HO3S-1 that lasted in service until 1957. It was followed by the HO4S utility helicopter, which the US Marine Corps designated the HRS. All HO4S variants in American military service became the H-19 series in 1962 with 1,000 units built for the American military.

The Sikorsky H-19 series was replaced in US Navy service in 1953 by a lengthened and more powerful version of the helicopter that was designated the HSS Seabat. It was configured for the ASW role. Another version that went to the US Marine Corps was labelled the HSS Seahorse and served in the utility and troop transport role. In 1962, the HSS Seabat became the SH-34 series with 382 units being built for the US Navy while the HSS Seahorse became the UH-34 series.

The early 1970s replacement for the HSS Seabat in the ASW role was the Kaman twin-engine SH-2 Seasprite. The helicopter started its service life with the US Navy as a single-engine utility helicopter in 1962, and was then designated UH-2. However, the demand for an ASW helicopter that could fly off the US Navy's non-carrier ships resulted in the service having the UH-2 rebuilt for the new role. One hundred and eighty-four units of the Seasprite, in various models, were built and the ASW version lasted in US Navy service until 2001. Kaman also supplied the US Navy twenty-four units of a utility helicopter designated the HUK-1 in the late 1950s.

The first-generation Sikorsky helicopters would be gone from US Navy use by the early 1970s. In their place came the Sikorsky twin-engine helicopter named the Sea King. Entering service with the US Navy in 1964, it was primarily an ASW helicopter that could also perform a variety of other roles such as surface warfare, SAR, aerial mine-sweeping, transport, and general utility duties. This is reflected in the various models of the helicopter built that would remain in US Navy service until 2006.

Entering into service the same year as the Sikorsky Sea King with the US Navy was the Vertol Sea Knight transport helicopter. Piasecki had become Vertol in 1955. Like the previous helicopters the US Navy had acquired from Piasecki, the Sea Knight was a tandem rotor helicopter. However, unlike its earlier cousins it was powered by two engines and not one.

The initial version of the Sea Knight acquired by the US Navy was designated the UH-46A. It remained in US Navy service until 2004. The bulk of the over 500 units of the Sea Knight constructed went to the US Marine Corps, with the last planned to be retired in 2015.

Initially acquired by the US Marine Corps as a twin-engine heavy transport heli-copter was the Sikorsky Sea Stallion, designated the CH-53A. The US Navy acquired fifteen from the US Marine Corps in 1971 for the aerial mine-sweeping role and labelled them the RH-53A. In 1973, the US Navy took into its inventory thirty units of a more advanced version of the twin-engine Sea Stallion, configured as an aerial mine-sweeper and designated the RH-53D.

Because the services wanted a more powerful version of the Sea Stallion, Sikorsky came up with an enlarged version with three engines instead of two, which the US Navy and US Marine Corps took into service in 1980 as the CH-53E Super Stallion. A total of 177 units of the Super Stallion were built. In 1986, the US Navy acquired the first of forty-six units of a specialized aerial mine-sweeping model of the Super Stallion, referred to as the MH-53E Sea Dragon.

Since the early 1980s, the bulk of the US Navy's helicopter inventory has been made up of a large variety of different models of the Sikorsky SH-60 Seahawk series, which was based on the US Army's UH-60 Blackhawk series helicopter also designed and built by Sikorsky.

The first of the SH-60 Seahawk series to enter US Navy service in 1984 was the SH-60B configured for the ASW and anti-surface role. It flew off frigates, destroyers, and cruisers. It replaced the Kaman Seasprite SH-2 ASW helicopter. A version of the Seahawk, labelled the HH-60H, was optimized for combat search and rescue (CSAR) and naval special warfare (NSW) missions and also appeared in the early 1980s.

Another version of the Seahawk was designated the SH-60F and intended for operation off carriers with the same job description as the SH-60B. It was the replacement for the Sikorsky SH-3 Sea King.

(*Above*) In a cloud of smoke generated by JATO rocket boosters, a nuclear-capable, prop-driven, Lockheed P2V-3C Neptune vaults off a carrier flight deck. As the senior leadership of the US Navy would not allow the Neptune to be recovered on carriers, because it was too large, if war had come, the Neptune pilots would be flying one-way missions. (*RWP*)

(*Opposite above*) The first nuclear-capable US Navy aircraft that could be both launched and recovered on carrier flight decks was the North American AJ Savage. One of three prototypes of the aircraft is seen here in its natural metal finish. The first prototype flight of the Savage took place on 3 July 1948. (*NNAM*)

(*Opposite below*) On the flight deck of a US Navy carrier shortly before launching is a North American production AJ-1 Savage. It was not a popular aircraft because it was unreliable in service, since it had been rushed into production. Another problem was its size, which meant fewer aircraft could be carried onboard a carrier when a Savage squadron was assigned. (*NNAM*)

(*Above*) Due to its size and large internal bomb bay, the Douglas A-3 Skywarrior proved a suitable candidate for conversion for many other roles during its time in service with the US Navy. Here we see the aircraft in the photo-reconnaissance configuration with all the cameras it would be able to carry on a mission arrayed in front of it. In this role, the Skywarrior was labelled the A3D-2P up until 1962, when it became the RA-3B. (*DOD*)

(*Opposite above*) Pictured is the nuclear-capable replacement for the AJ Savage series landing onboard a US Navy carrier: the Douglas A-3 Skywarrior. The aircraft was even larger and heavier than its predecessor, and earned the nickname the 'Whale' by those who flew and maintained it. It had an internal bomb bay that could carry 12,000 lbs of ordnance. (*NNAM*)

(*Opposite below*) Another large jet-powered nuclear-capable bomber adopted by the US Navy was the North American A3J Vigilante. In 1962, it became the A-5 Vigilante. With its new designation came a new role as a photo-reconnaissance plane with the designation RA-5C, an example of which is shown here. It was widely used during the Vietnam War. (*DOD*)

(*Above*) The US Navy was very excited about the prototype testing of what eventually became the McDonnell FH-1 Phantom. This resulted in an order in March 1945 for a larger and more powerful version of the Phantom that would become the McDonnell F2H Banshee. The initial model was labelled the F2H-1, soon followed by the F2H-2, an example of which is seen here on a carrier elevator. (*NNAM*)

(*Opposite above*) Coming in for a landing on a carrier flight deck with its arrestor hook down, is a twin-engine, one-man, McDonnell FH-1 Phantom fighter. Armament on the plane consisted of four .50 caliber machine guns in its nose with provisions for hard points under its wing for air-to-surface rockets. (*NNAM*)

(*Opposite below*) Pictured is a single-engine, single-seat, North American FJ-1 Fury being pushed into the hangar deck of a US Navy carrier from one of the ship's elevators. The aircraft has a large circular air inlet in its nose. Armament consisted of six .50 caliber machine guns arranged around the air inlet. The Fury was the last US Navy carrier fighter to be armed with machine guns. (*NNAM*)

(*Above*) A non-starter for the US Navy was the Vought F6U Pirate, a single-engine, single-seat fighter seen here. Testing of the aircraft's prototypes revealed a number of serious design issues and the US Navy quickly realized it would never be suitable for carrier service. As a result, all of the aircraft ordered were soon relegated to secondary duties. (*NNAM*)

(*Opposite above*) The single-engine, single-seat Grumman F9F Panther series included the initial production versions designated F9F-2 up through the F9F-5. Pictured in 1953 is an example of an F9F-5 on a carrier flight deck. It had a maximum speed of almost 600 mph and a cruising speed of 481 mph. Armament consisted of four fuselage-mounted 20 mm automatic cannons and underwing hard points for a variety of ordnance. (*NNAM*)

(*Opposite below*) The Douglas F3D Skynight shown here was originally intended as an all-weather interceptor by the US Navy, but was eventually placed into production as a night fighter. The large twin-engine, two-seat aircraft had a radar mounted in its nose and was armed with four 20 mm automatic cannons and air-to-air missiles. The initial model was the F3D-1, followed by the F3D-2. (*NNAM*)

On the flight deck of a US Navy carrier is a Vought F7U Cutlass, a twin-engine, single-seat fighter. It was armed with four 20 mm automatic cannons and had provision for underwing hard points for a variety of armaments, including a single nuclear bomb. The first model of the aircraft was designated the F7U-1. The final production model was the much modified F7U-3. (*NNAM*)

Taking off from the flight deck of the *USS Intrepid* in 1954 is a Vought F7U Cutlass. The plane's odd shape was based on German research during the Second World War into high-performance tailless aircraft. Besides its four 20 mm automatic cannons, ninety-eight units of the F7U-3 were modified to carry the Sparrow I air-to-air missile. The missile-armed version of the aircraft was designated the F7U-3M. (*NNAM*)

Another US Navy carrier fighter there had been high hopes for, but proved a disappointment in the end, was the single-engine, single-seat, McDonnell F3H Demon, seen here. Originally only a night fighter, it was large and heavy, and plagued by the inability of the service to fit it with an engine of sufficient thrust to power it. *(NNAM)*

A McDonnell F3H Demon is shown here being launched from a US Navy carrier. The first production model of the Demon to fly off carriers was a re-engined model labelled the F3H-2, of which 239 were constructed, all armed with four 20 mm automatic cannons. Eighty units of the F3H-2 variant were equipped to carry four air-to-air Sparrow missiles and designated the F3H-2M. *(NNAM)*

The Douglas F4D Skyrays shown here were acquired by the US Navy and US Marine Corps as high-speed carrier interceptors. There was only a single model of the aircraft placed into series production, designated the F4D-1. It was armed with four 20 mm automatic cannons and could carry four air-to-air missiles. In 1962, the F4D-1 became the F-6A. (NNAM)

Having just landed on a carrier flight deck is a Grumman F9F-6 Cougar, a single-engine, single-seat fighter, in US Navy markings. The initial production version of this aircraft was designated the F9F-6, with 646 units constructed, followed by 168 units referred to as the F9F-7. A photo-reconnaissance version was labelled the F9F-6P. The final production model of the plane was labelled the F9F-8 of which 602 were built. (RWP)

Seen in flight during an airshow is a restored North American FJ-4B Fury is on display at an airshow, the last version of the aircraft built. The initial version of the single-engine, single-seat fighter was labelled the FJ-2, which was followed by the FJ-3, and finally the FJ-4. The final sub-variant of the FJ-4 was designated the FJ-4B, with 222 constructed. The Fury series was armed with four 20 mm automatic cannons. *(Christophe Vallier)*

The single engine, single-seat Grumman F11F-1 Tiger fighter seen here was designed and built around the same parameters as the Grumman F8F Bearcat prop-driven fighter – to cram the most powerful jet engine then available into the lightest and smallest airframe possible. The first and only production model was designated the F11F-1. *(NNAM)*

Being readied for take-off from the flight deck of a US Navy carrier is the single-engine, single-seat Vought F-8 Crusader. Like many of the US Navy jet fighters that came before it, the main armament consisted of four 20 mm automatic cannons. Later US Navy fighters depended on air-to-air missiles as their main armament. The final production model of the aircraft was the F-8E. (NNAM)

(Opposite page) Pictured is the first flight of the F4H-1F Phantom II in May 1958. It was the initial model of the Phantom II, and included forty-five pre-series production units all based on the design of the two prototypes of the aircraft, labelled the XF4H-1. The F4H-1F was designated the F-4A in 1962 and like all the models to follow was powered by two engines and had a two-man crew. (NNAM)

Following the F-4A Phantom II off the assembly line were 649 production units of the F-4B Phantom II model for the US Navy and US Marine Corps, a museum example is seen here. The first units were delivered in 1961 and the last in 1967. As the US Navy inventory of F-4Bs units began showing their age, it was decided in 1970 to rebuild and upgrade 228 units with the new designation F-4N.
(Christophe Vallier)

The second model of the F-4 Phantom II series taken into US Navy and US Marine Corps service was the improved F-4J, of which an example is seen here just before launching from the *USS America*. A total of 522 units were built between 1966 and 1972. Eventually, as this version of the aircraft aged, it was decided to rebuild and upgrade it in the late 1970s. The rebuilt and upgraded F-4J was assigned the new designation F-4S. *(DOD)*

One of the most interesting innovations of the failed General Dynamics F-111B fighter, that was considered but rejected by the US Navy as a replacement for the F-4 Phantom II series, was the adoption of variable geometry (moving) wings. In this picture of a prototype F-111B, the wings are swept back allowing the aircraft to reach its maximum speed. The plane had a two-man crew and was powered by two engines. *(NNAM)*

Parked next to an F-14A Tomcat is a two-seat trainer version of a Soviet-designed MiG-15 fighter that entered service in 1949, of which 18,000 were eventually built in various versions. Four hundred and seventy-eight units of the F-14A were built. The follow-on F-14B model consisted of thirty-eight brand new units and forty-eight F-14As brought up to the B model standard. *(DOD)*

This photograph of a Grumman F-14A Tomcat shows its variable-geometry wings swept forward for maximum lift prior to being launched. With its wings spread the aircraft had a wingspan of 64 feet, which dropped to 38 feet when swept back, which can be further reduced to 33 feet in a special 'over-swept' position for parking on the carrier's crowded deck. With the wings brought forward the aircraft's speed could be reduced enough to safely recover and launch it from carrier flight decks. *(DOD)*

Coming in for a landing with all its flaps down and its ailerons acting as 'flaperons' is a US Navy McDonnell Douglas-Northrop F/A-18A Hornet, the initial series production model built. The aircraft was intended by the US Navy to be a multi-role plane that was more affordable to buy and fly than the F-14 Tomcat series. (*DOD*)

Pictured is the cockpit of an early model McDonnell Douglas-Northrop F/A-18 Hornet with the pilot's multi-function displays shown. All the various models of the Hornet series carry a 20 mm automatic cannon internally. In the interceptor role, the aircraft is armed with externally-mounted air-to-air missiles. In the strike role, the Hornet can carry 13,700 lbs of ordnance and external fuel. (*DOD*)

The Douglas Skyraider series of prop-driven attack planes encompassed a great many models. Pictured is a restored AD-4 Skyraider in the paint scheme and markings of a US Navy aircraft that saw service during the Korean War. There were 372 units of the AD-4 version built for the US Navy and US Marine Corps. In 1952 it was modified to deliver nuclear bombs. (*Christophe Vallier*)

Pictured is a restored Douglas A-1H Skyraider in a US Navy Vietnam War paint scheme and markings. Prior to 1962, it was designated the AD-6. There were 713 units of this model built. Beginning with the previous model, the AD-5, which was re-labelled the A-1E in 1962, the Skyraider was built as a basic model that, depending on mission requirements, could be converted for a number of different roles. (*NNAM*)

One of the competitors for the attack role performed by the Douglas Skyraider for the US Navy was the Martin AM-1 Mauler seen here. The Mauler turned out to be heavier than the Skyraider and its operational parameters were inferior. Originally, the US Navy had ordered 750 units of the Mauler in early 1945, but design problems with the aircraft resulted in the production order being cancelled in October 1949, with the 144th plane built. (NNAM)

On display at an air show is a restored example of a US Navy Douglas A4D-2N Skyhawk. A total of 638 units of this aircraft were constructed. In 1962 the aircraft was re-designated the A-4C. Unlike the earlier versions of the Skyhawk that were strictly daytime fair-weather aircraft, the A-4C model was fitted with additional avionics that allowed it to operate at night and in adverse weather conditions. (Christophe Vallier)

In flight is a US Navy Douglas A-4F Skyhawk, with the very noticeable add-on 'hump back' pod that contained additional avionics. One hundred and forty-seven units of the A-4F were built, with deliveries beginning in 1967. Eventually, all 499 units built of the previous A-4E model were also fitted with the add-on avionics pod. (*DOD*)

Pictured is a US Navy Vought A-7E Corsair light attack aircraft intended only for daylight, clear weather operations. It was a single-engine single-seat aircraft armed with a single 20 mm automatic cannon, and could carry an ordnance load of 15,000 lbs. The plane's maximum speed was 698 mph. It first saw combat in the Vietnam War in December 1967. (*DOD*)

Pictured is the first Grumman A-6A delivered to the US Navy. It was a twin-engine, two-seat all weather/night dedicated attack aircraft. It had five hard points capable of hauling into battle 18,000 lbs of ordnance. Four hundred and eighty-four units of the A-6A variant were built; nineteen were later converted to the A-6B version during the Vietnam War that specialized in attacking enemy air defense units. *(NNAM)*

(Opposite page) The final model of the Grumman A-6 Intruder series was designated the A-6E and an example is seen here dropping its bomb load during a training exercise. A total of 480 units of the A-6E entered into US Navy and US Marine Corps service. 240 were newly built, with the remaining 240 converted from earlier models of the aircraft. *(NNAM)*

On the flight deck of a US Navy carrier is one of the sub-variants of the Grumman A-6 series, referred to as the EA-6B Prowler, of which 170 units were built. The EA-6B has a stretched airframe to allow it to accommodate a crew of four, as is evident by the second cockpit seen in this picture. *(NNAM)*

The Martin XP6M-1 prototype SeaMaster seaplane seen here was the US Navy's fallback nuclear-capable strategic bomber if it failed to acquire the super carriers it needed that could handle the large and heavy wheeled nuclear-capable strategic bombers then being planned for. The SeaMaster had a number of unresolved design issues. By the time the technical problems were resolved, the US Navy lost interest in the program and it was cancelled. (*NNAM*)

(*Opposite page*) An artist's impression of the never-built General Dynamics/McDonnell-Douglas A-12 Avenger all weather/night dedicated attack aircraft that was the intended replacement for the Grumman A-6 Intruder series. Power for the two-man aircraft was to be provided by two engines. (*NNAM*)

Pictured during one of its test flights is the prototype Convair XF2Y-1 Sea Dart seaplane fighter. The single-engine, single-seat aircraft rode across the water on skis until taking flight. By 1955, it was clear to the US Navy that a newer generation of supersonic carrier-capable fighters would be far superior to the Sea Dart and the program was ended. (*NNAM*)

Coming in for a landing during an airshow is a restored Grumman HU-16 Albatross seaplane in US Navy markings. The aircraft was manned by a crew of between four and six men and could carry ten passengers. It had a wingspan of 98 feet 8 inches. (*Christophe Vallier*)

The Martin Marlin seaplane had a re-designed fuselage and more powerful engines compared to its forerunner, the Martin Mariner seaplane. In US Navy service there was the P5M-1 Marlin and the modified P5M-2 Marlin. In 1962, the P5M-1 became the P-5A and the P5M-2 became the P-5B. The P-5B model of the Marlin had a T-shaped tail as seen in this picture of one in US Coast Guard markings. (*NNAM*)

On museum display is a Lockheed land-based Neptune maritime patrol plane, designated the SP-2H, the last model of the aircraft that entered service in 1965. A total of 287 units of the SP-2H version of the plane were built and a number of sub-variants appeared to perform a variety of roles. The letter 'S' that begins the aircraft designation code is a mission modification letter that stands for antisubmarine. (NNAM)

Losing out to the Neptune as the US Navy's preferred choice as a land-based maritime patrol aircraft was the Martin P4M Mercator, of which twenty-one were built for the US Navy. As with the SP-2 it was powered by two prop-driven engines. Not wanting to waste the Mercator units built, eighteen were modified for a new job in 1949. That new role had them converted into electronic intelligence (ELINT) aircraft, and assigned the designation P4M-1Q. (NNAM)

The eleven-man P-3C Orion seen here can be armed with up to 15,000 lbs of ordnance, including air-to-surface missiles and rockets, as well as mines, torpedoes, and nuclear-armed depth charges. This ordnance is carried either in an internal bomb bay or ten underwing hard points. The long object protruding from the rear of the aircraft fuselage is the detector for the Magnetic Anomaly Detection (MAD) gear intended to pick up the magnetic signature of submarines. (DOD)

The Grumman AF Guardian was the first dedicated US Navy anti-submarine warfare (ASW) aircraft employed post-war. The Guardian seen here with the large under-fuselage radome was the detection aircraft that worked together with a weapon-armed Guardian when hunting submarines. It proved to be the largest and heaviest single-engine prop-driven aircraft to be launched and recovered from US Navy carriers. (NNAM)

This photograph shows an early model Grumman S2F-1 Tracker employed by the US Navy as a carrier-capable anti-submarine warfare aircraft. In its internal bomb bay there was space for two anti-submarine torpedoes or a single nuclear depth charge. The radome on the top of the aircraft's fuselage, just behind the cockpit, was for an Electronic Support Measure (ESM) pod that detected radar transmissions. (*NNAM*)

Derived from the Grumman S2F-1 Tracker anti-submarine warfare (ASW) aircraft was the Grumman C-1 Trader, shown here on a carrier flight deck. The aircraft became the TF-1 in 1962 with eighty-three units built. The US Navy referred to this type of aircraft as a Carrier Onboard Delivery (COD) plane. (*DOD*)

Being hoisted aboard the *USS Kitty Hawk* in 1977 is a Lockheed S-3A Viking originally configured as an anti-submarine warfare (ASW) aircraft. As an ASW plane it had a crew of four. When the airplane was reconfigured for the anti-surface warfare (ASUW) role, all the submarine detection gear was removed and the crew was reduced to only two. (*NNAM*)

The US Navy PO-1W Warning Star seen here was an airborne early warning (AEW) aircraft based on the Lockheed Constellation passenger plane. On top of the aircraft's fuselage is a large vertical radome for a height-finding radar antenna and on the bottom of the fuselage is a radome to house an air-search radar antenna. The typical mission crew for the aircraft was eighteen men. (*RWP*)

The two aircraft in this picture are both postwar-developed US Navy airborne early warning (AEW) planes that were carrier-capable. In the background is the older of the two, the WF-2 Tracer, which was re-labelled the E-1B Tracer in 1962. It had a mission crew of four, including two pilots and two radar operators. The aircraft in the foreground is the Tracer's replacement, the larger Grumman E-2A Hawkeye. (NNAM)

On the fuselage roof of the Grumman E-2A Hawkeye shown here is the large 24-foot diameter radome that housed the antenna for the aircraft's search radar unit. The mission crew of the Hawkeye consists of two pilots and three radar operators. (NNAM)

Based on the Hawkeye-series of airborne early warning (AEW) aircraft is the Grumman C-2A Greyhound Carrier Onboard Delivery (COD) airplane seen here moments before being launching from a US Navy carrier flight deck. It first showed up in US Navy service in 1965 and continues to serve the delivery needs of carriers. It was the replacement for the TF-1 Trader, with fifty-eight units built. (*DOD*)

The prop-driven aircraft with the longest wingspan that remains in the US Navy inventory is the Lockheed C-130 Hercules series. An example is seen here in the markings of the Navy's well-known Blue Angels Flight Demonstration Team. Other units of the Hercules see employment in the US Navy and US Marine Corps as both logistic and air refuelling aircraft. (*DOD*)

A training aircraft no longer in service with the US Navy is this preserved example of a North American T-2B Buckeye in US Navy training command colours and markings at an airshow. The two-seat aircraft was powered by two engines and had a maximum speed of 522 mph. It was carrier-capable and served with the US Navy from 1959 until 2008, with a total of 529 units built. *(Christophe Vallier)*

The replacement training aircraft for the North American T-2C Buckeye and the Douglas TA-4J Skyhawk in US Navy service was the two-seat British Aerospace T-45A Goshawk seen here, which remains in service today. It first showed up in US Navy service in December 1991. It has a maximum speed of 609 mph and is carrier-capable as is evident by the arresting hook seen on the bottom of the rear fuselage. *(DOD)*

To help US Navy and US Marine Corps fighter pilots hone their combat skills by making simulated combat missions as realistic as possible, a number of aircraft were acquired that were dissimilar in appearance to standard naval aircraft and would replicate the performance of potential enemy fighters. One of those aggressor aircraft types adopted was the Northrop F-5E, seen here. (DOD)

Another dissimilar aircraft acquired by the US Navy for help in training its fighter pilots, as well as those of the US Marine Corps, at the Naval Strike and Air Warfare Center (NSAWC) is the General Dynamics F-16 Fighting Falcon series, an example of which is seen here. At first, the US Navy flew twenty-six modified examples of the C model of the aircraft. These were retired in 1991, and replaced by fourteen upgraded examples of the A and B versions of the Fighting Falcon. (Christophe Vallier)

The three Piasecki HRP-I Rescuer helicopters seen here in US Marine Corps markings were acquired between 1947 and 1949 by the US Navy, who quickly decided to pass most of them off to the other services. The helicopter had a two-man crew and could carry up to ten passengers, or 2,000 lbs of cargo. (*NNAM*)

On museum display is a Piasecki HUP-2 Retriever. Like the HRP-I Rescuer helicopter that came before it, the HUP-2 was a tandem-rotor design powered by a single engine. The Retriever was 32 feet long and had a rotor diameter of 35 feet. With a two-man crew, the Rescuer could carry up to five passengers. (*Christophe Vallier*)

Landing on a moving ship is always dangerous, as is evident in this picture of a crashed helicopter, referred to by the US Navy as the HO3S-1. This helicopter performed a very useful role in rescuing American military aircrews downed behind enemy lines during the Korean War. It had a single pilot and seating for three passengers. (*National Archives*)

Also seeing use with the US Navy and US Marine Corps during the Korean War was the replacement for the Siko HO3S-1; the Sikorsky company-designated S-55. In US Marine Corps service, as seen here during the Korean the helicopter was labelled the HRS-1 through HRS-3, and in US Navy service as the HO4S-1 and HO-(*National Archives*)

In need of a dedicated anti-submarine warfare (ASW) helicopter the US Navy took into service another Sikorsky helicopter, known by its company designation as the S-58. In US Navy service it was labelled the HSS-1 Seabat and could be armed with a variety of depth charges, mines, or homing torpedoes. Many were used in the utility role by the US Navy. (NNAM)

The Kaman SH-2F Seasprite shown here was intended by the US Navy as a ship-based anti-submarine warfare helicopter that was also able to engage small hostile surface ships. The helicopter crew consisted of a single pilot, a tactical control officer and a sensor operator. It had a maximum speed of 165 mph. (DOD)

The Sikorsky Sea King helicopters seen here were mainly intended for the anti-submarine warfare role operated from US Navy carriers, unlike the smaller Kaman SH-2F Seasprite that operated from non-carriers. The initial model of the Sea King was designated the SH-3A with 245 units built. A later, more advanced version was labelled the SH-3D with 73 built and two converted from the earlier SH-3A variant. (DOD)

The Vertrol UH-46A was the first model of the Sea Knight helicopter acquired by the US Navy. It had a two-man crew and could carry seventeen passengers or 4,000 lbs of cargo. There was also a specialized search and rescue (SAR) model of the Sea Knight developed for the US Navy with the initial model labelled the HH-46A and a follow-on version as the HH-46D. (DOD)

Shown in flight is a twin-engine Sikorsky Sea Stallion helicopter configured for the aerial mine-sweeping role by the US Navy during the Vietnam War and designated the RH-53D. Notice the in-flight refuelling probe projecting from the front of the helicopter's fuselage. With the introduction of the three-engine Sikorsky MH-53E Sea Dragon into US Navy service, the RH-53D units remaining in US Navy service were converted back into transport helicopters. (*DOD*)

The Sikorsky MH-53E Sea Dragon pictured will remain in US Navy service until at least 2025. It can be distinguished from the CH-53E Super Stallion by the large sponsons seen here on either side of the fuselage that contain more fuel storage to increase the range of the helicopter. It is flown by two pilots and has a crew of between one to six people, depending on the mission. (*DOD*)

In flight is a Sikorsky SH-60B Seahawk helicopter, also referred to as the (LAMPS) Light Airborne Multipurpose System Mk. III. It was the replacement for the Kaman SH-2 Seasprite on US Navy warships other than carriers. It was designed to be able to both engage below- and above-surface threats in all weather conditions day and night. Like almost all US Navy helicopters, it can also perform many secondary roles. (*DOD*)

Chapter Four

Post Cold War Aircraft

In the aftermath of the long-running Cold War, which ended in 1991 with the collapse of the former Soviet Union and the disintegration of its once powerful military machine, the US Navy's budget, along with the other armed services of the United States, was dramatically reduced. This resulted in some naval aircraft eventually being phased out of service, updates to others being cancelled, and some being pushed into new roles.

The F-14 Tomcat Soldiers On

In 1991, the final model of the F-14 Tomcat series appeared, labelled the 'D' version. A proposed 'C' version was never funded. Initial plans made in 1984 by the US Navy had envisioned the production of 324 new units of the F-14D Tomcat. When that plan was dashed due to funding shortfalls, the US Navy's backup plan (conceived in 1989) called for upgrading of 400 units of the original F-14A Tomcat to the 'D' model standard. This plan also foundered for lack of funding, and what the US Navy eventually got was only thirty-seven brand new F-14D Tomcats and eighteen units of the F-14A upgraded to the 'D' model standard, with the last unit being delivered in 1992.

As the threat of Soviet aerial attacks on US Navy carrier battle groups disappeared following the Cold War, the need for a dedicated carrier fighter such as the F-14 Tomcat series was minimal. However, with the pending withdrawal from service of its aging inventory of Grumman A-6 Intruder strike aircraft, beginning in the 1990s, and a new version of the F/A-18 Hornet series intended to assume that role not yet in production, the US Navy decided to reconfigure the F-14 Tomcat series for the strike role. In this new role the aircraft was unofficially nicknamed the 'Bombcat'.

For the strike role the Tomcat series was cleared to drop unguided (dumb) bombs in 1992, but more precision was demanded, so in 1994 it was provided with the targeting pod from the Low Altitude Navigation and Targeting Infrared for Night (LANTIRN) system to allow use of guided (smart) bombs. The Tomcat series had also acquired the ability to perform the photo-reconnaissance role with the addition of TARPS (Tactical Air Reconnaissance Pod System). Despite the addition of TARPS and LANTIRN pods, the Tomcat series was retired from US Navy service in 2006, with the F-14A model going first in 2004.

A New Model Hornet

One of the key additions to the US Navy's post-Cold War inventory of carrier aircraft was the acquisition of two advanced models of the Boeing F/A-18 Hornet series, which includes the single-seat F/A-18E and the two-seat F/A-18F versions. Unofficially these two newest models of the F/A-18 Hornet series are often referred to as 'Super Hornets', and this name will be used for the remainder of the text.

The F/A-18 Super Hornets began showing up in carrier service in 2001 replacing the F-14 Tomcat series. Reasons for the replacement of the F-14 Tomcat series include its poor in-service reliability record and the belief by some that it was obsolete.

The F/A-18 Super Hornet is larger and heavier than the original F/A-18 Hornet series. The first generation of F/A-18 Hornets are now unofficially referred to as the 'Legacy Hornets' or 'Rhinos' on carriers, to distinguish them from the new F/A-18 Super Hornet models. Due to the increase in size, the F/A-18 Super Hornet series carries a great deal more fuel than the original F/A-18 Hornet series aircraft, which were criticized by many in the US Navy as being too short-ranged in comparison to the Grumman F-14 Tomcat and A-6 Intruder.

The US Navy currently believes that having a single common platform aircraft with modern avionics performing a number of different roles on carriers is much more cost-efficient than having a wide variety of different aircraft performing those roles. The F/A-18 Super Hornet was therefore designed from the beginning to be reconfigured for performing many different jobs. Some of them include air dominance, fighter escort, all weather day/night precision strike, and reconnaissance.

One of the roles assigned to the F/A-18 Super Hornet was the ability to be configured as a tanker when required. In such cases the aircraft is fitted with a removable aerial refuelling system (ARS). This was a much welcomed addition to the US Navy's carriers as it took over the role once performed by the aging Grumman KA-6D Intruder tanker variant and the Lockheed S-3 Viking when configured as a tanker.

A dedicated EW version of the F/A-18F Super Hornet intended for the air defense suppression role is the EA-18G Growler that first showed up on US Navy carriers in 2009 It replaced the Grumman EA-6B Prowler, the EW version of the Grumman A-6 Intruder, that will be pulled from US Navy service in 2015. The EA-18G Growler is also tasked with performing the EW for the US Air Force if called upon.

With the ever increasing speed at which technology evolves, the US Navy is in a near constant process of upgrading all its aircraft, the Super Hornet being no exception. Since its introduction into service, the aircraft has already gone through two upgrade programs, which in current military terminology are referred to as 'Block I' and 'Block II'. The US Navy is also considering several additional upgrade programs for the aircraft in the future.

In 2015, the US Navy had approximately 560 Super Hornets in its inventory. The original plans had called for a total of 1,000 to be acquired but funding shortfalls prevented this from happening.

The Fifth-Generation Fighter

The replacement for the US Navy's inventory of F/A-18 Hornets and Super Hornets is one of three variants of the Lockheed Martin F-35 Lighting II multi-role fighter. The US Navy version is known as the F-35C, while the US Marine Corps variant is the F-35B. A US Air Force model is referred to as the F-35A. The original thought behind the F-35 Lighting II multi-role fighter was that a single aircraft type, with minor differences, could serve all three services, and result in significant cost savings for the American taxpayer.

Technically, according to the 1962 Tri-Service Aircraft Designation System mandate, the F-35 Lighting II multi-role fighter should have been labelled the F-24, but that did not happen, as the then Secretary of Defense for Acquisition, Training and Logistics decided, in a 2001 press conference, that it should be the F-35, because the experimental model of the aircraft was labelled the X-35.

The US Marine Corps F-35B is the replacement for their inventory of legacy F/A-18 Hornets and McDonnell-Douglas (now Boeing) AV-8B Harrier light attack aircraft. Unlike the other two variants of the F-35, the US Marine Corps version has a short take-off and vertical landing (STOVL) capability, as does the AV-8B Harrier. The US Marine Corps will also acquire a number of the US Navy's F-35C version for use on carriers.

The US Navy's F-35C has folded wing tips and a strengthened undercarriage to absorb the stresses of being recovered on board carriers. To this end, it has a twin-wheel nose gear, rather than the single nose wheel seen on the other two variants of the aircraft. In addition, the F-35C has larger wings and tail compared to the other versions of the aircraft, to improve low-speed control of the aircraft when approaching a carrier flight deck. An added benefit of larger wings for the F-35C is an increase in range and payload.

Unmanned Drones

An unmanned drone already operating off non-carrier ships is referred to as the Fire Scout, and has appeared in two versions so far, MQ-8B and the MQ-8C. The MQ-8B and MQ-8C Fire Scouts are modified helicopters, so they don't need a large flight deck to operate. Both of the Fire Scouts are classified by the US Navy as unmanned air systems (UAS). The MQ-8B was first deployed in 2011, and the larger MQ-8C's first flight was 31 October 2013, with planned deployment in 2015.

A land-based unmanned drone planned to be employed by the US Navy is a Northrop-Grumman product designated the MQ-4C Triton. Unlike the proposed

UCLASS system, the Triton is only intended as a long-range, high endurance surveillance aircraft and will not be configured to carry weapons. The first prototype of the Triton was delivered to the US Navy in 2012 for testing. The US Navy hopes to have it in service in 2017.

The planned US Navy replacement for the F/A-18 Super Hornet series in the future to serve alongside the F-35C Lighting II onboard carriers will be what is currently referred to as an unmanned combat air vehicle (UCAV). A test unit, designed and built by Northrop-Grumman, designated the X-47B, successfully flew off and landed on a carrier in 2013. It was also tested to see how it would integrate with personnel and manned planes on a carrier flight deck. Testing of the X-47B will continue until at least 2015. There is a proposed larger version, to be known as the X-47C.

If testing of the X-47 series prototypes is judged to be successful, the US Navy plans on having Northrop-Grumman develop a combat-ready version that is currently being referred to as the Unmanned Carrier-Launched Surveillance and Strike (UCLASS) system. If all goes well, the US Navy is hoping to have the weapon-equipped unmanned drone in carrier service by 2020, although that is optimistic.

New Role

With the end of the Cold War in 1991, and the submarine threat disappearing, the US Navy's Grumman S-3 Viking had its submarine detection gear removed. It was then optimized for the surface warfare threat by being configured to carry a number of air-to-surface missiles. Another role it was configured for was aerial tanker, with the addition of external fuel tanks. The last S-3 Viking was pulled from service in 2009.

Sixteen units of the S-3 Viking series were converted into ELINT aircraft in 1993 and re-designated as the ES-3A Shadow. They were the replacement for the Douglas EA-3B Skywarrior. They lasted in service until 1999, when they themselves were replaced by a next generation aircraft.

New Maritime Patrol Aircraft

The US Navy's replacement for its aging inventory of P-3 Orion maritime patrol aircraft is the Boeing P-8A Poseidon. It first entered service in 2013 and is based on a militarized version of the Boeing 737-800 civilian passenger plane, with components borrowed from other 737 series aircraft. Unlike its predecessor that was prop-driven, the P-8A Poseidon is jet powered. Besides the anti-submarine weapons carried onboard, the aircraft also is able to launch air-to-surface missiles to engage hostile surface ships. It is intended that the P-8A Poseidon will work in conjunction with the MQ-4C Triton unmanned drone when introduced into service.

Updates for Old Favorites

Among the Cold War-era US Navy aircraft that continue to serve in the post-Cold War era are upgraded versions of the Northrop Grumman E-2 Hawkeye series, and its single variant the Grumman C-2A Greyhound, of which the US Navy currently has thirty-five units in service, sometimes referred to as the C-2A(R). These went through a Service-Life Extension Program (SLEP) in 2011.

The current version of the E-2 Hawkeye series is the C model that entered into service in 1973 and has been through a number of upgrades up into 2014. In 2013, the aircraft was fitted with all-composite eight-bladed propellers. In total, fifty-five units of the E-2C Hawkeye entered into service. The US Navy replacement for the E-2C Hawkeye is the Northrop Grumman E-2D Advanced Hawkeye, of which the US Navy plans on acquiring seventy-five examples.

Helicopters

The first post-Cold War-era Sikorsky SH-60 Seahawk-series helicopter acquired by the US Navy was the MH-60R in 2002. It combined the features and roles of the earlier SH-60B and SH-60F Seahawk helicopters into one platform. Reflecting its many jobs, the US Navy assigned the mission modification code letter 'M' at the beginning of the aircraft's letter prefix code to denote it being a multi-mission platform. Armed with ASW weapons as well as air-to-surface missiles, the MH-60R Seahawk can handle a wide array of threats. All US Navy units equipped with the MH-60R Seahawk are known as Helicopter Maritime Strike Squadrons.

Besides the MH-60R, the US Navy took another version of the SH-60 Seahawk into service about the same time, designated the MH-60S Seahawk. It was the replacement for the US Navy's inventory of CH-46 Sea Knight helicopters. In honour of its predecessor, the MH-60S Seahawk is unofficially sometimes referred to as the Knighthawk.

Unlike the single mission CH-46 Sea Knight, the MH-60S Seahawk can be configured to perform a wide variety of combat and non-combat missions. It can be armed with a number of different types of machine guns as well as ASW weapons and air-to-surface missiles. US Navy units equipped with the helicopter are known as Helicopter Sea Combat Squadrons. The US Navy phased out of service in 2015 all the earlier versions of the SH-60 Seahawk, with the MH-60R and MH-60S Seahawks to assume all their roles.

(*Above*) The last version of the Grumman Tomcat series authorized was the F-14D, seen here being launched. It came with a variety of upgraded avionics, including a new AN/APG-71 radar that offered a dramatic improvement in performance over the AWG-9 radar on earlier generation Tomcats. The AWG-9 radar was an analog device inherited from the failed F-111B program rejected by the US Navy, whereas the AN/APG-71 radar is an all-digital device that can both search and track a large number of potential enemy aerial threats at the same time. (*DOD*)

(*Opposite above*) Empty, the Grumman F-14D Tomcat seen here weighed 41,780 lbs. With a full load of fuel and an assortment of ordnance it could weigh up to 74,000 lbs, when catapulted off a carrier flight deck. The aircraft shown has two large external fuel tanks, one mounted under each of the large air intakes, and an assortment of ordnance mounted under the bottom of its fuselage. (*DOD*)

(*Opposite below*) Shown coming in for a landing on a US Navy carrier flight deck is a Grumman F-14D Tomcat. In the foreground are the ship's landing signal officers who assist pilots in landing by providing input via radio on their approach to a ship's flight deck. With variable-geometry wings, the F-14D had a wingspan of 64 feet 1½ inches in the forward-swept position and 38 feet 2½ inches in its swept configuration. There is also an over-sweep position for parking the aircraft. (*DOD*)

(*Above*) Pictured is the two-seat version of the Boeing Super Hornet, designated the F/A-18F. It and the single-seat Super Hornet come with a radar that can be employed in the air-to-air search mode, as well as a ground moving target indicator. In February 2014, the US Navy began testing a new external passive infra-red search and track sensor pod for the Super Hornet, to enhance it abilities to perform strike, reconnaissance, and surveillance missions. (*DOD*)

(*Opposite above*) The single seat US Navy Boeing F/A-18E Super Hornet shown here is an improved and larger version of the original F/A-18 Hornet series, now referred to unofficially as the 'Legacy Hornet'. The Super Hornet is 60 feet in length and has a wingspan of 44 feet 8½ inches when air-to-air missiles are fitted to the tips of the plane's wings. The height of the aircraft is 15 feet 10 inches. (*DOD*)

(*Opposite below*) Being recovered on a US Navy carrier flight deck is a Boeing F/A-18E Super Hornet. Among the advantages bestowed upon the pilots flying the Super Hornet is the fact that it is approximately 11 mph slower on landing approach than the Legacy Hornet, offering pilots an extra margin of safety in the difficult job of carrier recovery. In the air the Super Hornet enjoys a greater degree of manoeuvrability than its earlier versions, an important feature in air-to-air combat. (*DOD*)

Among its many roles, the Boeing F/A-18 Super Hornet series can be quickly converted into an aerial refuelling aircraft by the addition of an external 330 gallon in-flight refuelling pod under its fuselage and four large 480 gallon underwing external fuel tanks, as shown with the F/A-18F pictured. (DOD)

Pictured is a version of the Boeing F/A-18 Super Hornet series, designated the EA-18G Growler. It is a variation of the two-seat F/A-18F Super Hornet and intended for the electronic warfare role. Unlike the Legacy Hornet and the normal Super Hornet, the aircraft is not fitted with an internal 20 mm automatic cannon. It can be armed with air-to-surface missiles optimized to destroy enemy air-defense sites. (DOD)

The planned replacement for the early model Hornets onboard US Navy carriers is the single-seat Lockheed Martin F-35C Lightning II seen here. US Navy plans call for the eventual acquisition of 340 of the aircraft, of which 80 will be operated off carriers by the US Marine Corps. Instead of a heads-up display, the Lightning II pilot has a fully integrated helmet-mounted display system (HMDS). *(DOD)*

Unlike the early model Hornets and Super Hornets powered by two engines, the Lockheed Martin F-35C Lightning II shown here has only a single engine. It does not have an internal automatic cannon as the Hornet series, rather an external pod containing a 25 mm multi-barrel automatic cannon and 182 rounds of ammunition can be fitted if required. The US Air Force version of the Lightning II series, designated the F-35A, is equipped with an internal 25 mm cannon. *(DOD)*

This Northrop Grumman X-47B unmanned combat air vehicle is being catapulted from a US Navy carrier flight deck. Empty, the unmanned drone weighs 14,000 lbs. Maximum take-off weight is 44,567 lbs. It is powered by a single engine. As built, the X-47B has two internal weapon bays that can carry up to 4,500 lbs of ordnance. (DOD)

(Opposite page) Shown on one of the deck edge elevators of a US Navy carrier is the Northrop Grumman X-47B unmanned combat air vehicle that first flew in 2011. The original smaller proof-of-concept version of the unmanned drone was labelled the X-47A and was first flown in 2003. The X-47B has a length of 32 feet 2 inches and a wingspan of 62 feet 1 inch. (DOD)

Latched to the stern helicopter launching and recovery deck on a US Navy warship is the unmanned Northrop Grumman MQ-8B Fire Scout helicopter. It is based on the Schweizer (now Sikorsky) Model 333 four-person passenger helicopter. The MQ-8B Fire Scout can stay in the air for eight hours with a payload of 200 lbs. (DOD)

Unhappy with the small payload capacity of the unmanned Northrop Grumman MQ-8B Fire Scout helicopter, the US Navy sought out a larger helicopter. Northrop Grumman therefore chose the airframe of the Bell Helicopter Model 407 for the next version of the Fire Scout. It is seen here and referred to as the MQ-8C Fire Scout and has a payload capacity of 1,250 lbs, or a sling load of 2,650 lbs. (DOD)

Taking off from an airfield is a US Navy MQ-4C Triton unmanned drone, a variation of the US Air Force unmanned Global Hawk that first flew in 1998. The Global Hawk employed by the US Air Force took over the role once performed by the famous manned Lockheed U-2 spy plane. The first flight of the MQ-4C Triton for the US Navy took place in 2013. (DOD)

Shown are two US Navy MQ-4C Triton unmanned drones. The drones are intended for the maritime surveillance role and have a length of 47 feet 6 inches and a wingspan of 130 feet 9 inches. They weigh 32,250 lbs and have a service ceiling of 60,000 feet, with the ability to stay in the air up to 38 hours. (*DOD*)

Chained down to the flight deck of a US Navy carrier is a Lockheed ES-3A Shadow, based on the S-3 Viking series of ASW aircraft. From 1995 until 1999, the four-person Shadow performed the role of electronic surveillance aircraft, as well as over-the-horizon targeting for strike aircraft. It had a maximum endurance of seven hours in the air. (*DOD*)

Pictured is the replacement for the US Navy inventory of prop-driven Lockheed P-3C Orions, the jet-powered Boeing P-8A Poseidon. The US Navy envisions acquiring 109 units of the Poseidon. Among its many roles are antisubmarine warfare, anti-surface warfare, and electronic signals intelligence (ELINT). The aircraft has a typical crew of nine, two pilots and seven sensor operators. (DOD)

The Boeing P-8A Poseidon shown here is 129 feet 5 inches long and is 42 feet 1 inch in height. Wingspan is 123 feet 6 inches, making it larger than the Lockheed P-3C Orion and requiring the building of new hangars to house the aircraft. The aircraft is reportedly less expensive and time-consuming to maintain than its predecessor, the Lockheed P-3C Orion. (DOD)

In flight is a Northrop Grumman E-2C Hawkeye. The US Navy's inventory of the aircraft consisted of a combination of new-built planes, and older E-2Bs upgraded to the 'C' model standard. The radar in the E-2C Hawkeye can detect aircraft at a distance of 276 miles, and can track more than 250 aerial objects and direct up to 30 fighters in the interceptor role at the same time. *(DOD)*

The replacement for the Northrop Grumman E-2C Hawkeye is the Northrop Grumman E-2D Advanced Hawkeye, two examples of which are seen here. As with the updated E-2C Hawkeyes, the E-2D Advanced Hawkeye has all-composite eight-bladed propellers that produce less vibration and noise than the previous four-bladed metal propellers that date back to the introduction of the series. *(DOD)*

Coming in for a landing on a US Navy carrier flight deck is an upgraded Grumman C-2A Greyhound Carrier Onboard Delivery (COD) aircraft. The C-2A is based on the Northrop Grumman E-2 Hawkeye series and, as with the E-2C and E-3D models, the C-2As have been fitted with all-composite eight-bladed propellers. (*DOD*)

Pictured on the landing and recovery platform of a US Navy warship is a US Navy Sikorsky MH-60R Seahawk helicopter. The MH-60R is the replacement for the Sikorsky SH-60B and SH-60F in the anti-submarine warfare role. However, it is also capable of many other roles such as search and rescue, vertical replenishment, and anti-surface warfare. (*DOD*)

The US Navy Sikorsky MH-60S Knight Hawk seen here is both a non-combat logistical support helicopter, and in its latest version labelled the MH-60S Block IIIA/B, can be configured to carry a variety of armaments for different missions other than that of anti-submarine warfare. In 2015 the US Navy had 275 units of the MH-60S Knight Hawk in the inventory. (*DOD*)